T0210941

# Currants, Gooseberries, and Jostaberries
## *A Guide for Growers, Marketers, and Researchers in North America*

Danny L. Barney, PhD
Kim E. Hummer, PhD

## CRC Press
Taylor & Francis Group
Boca Raton  London  New York

CRC Press is an imprint of the
Taylor & Francis Group, an informa business

CRC Press
6000 Broken Sound Parkway, NW
Suite 300, Boca Raton, FL 33487
270 Madison Avenue
New York, NY 10016
2 Park Square, Milton Park
Abingdon, Oxon OX14 4RN, UK

For more information on this book or to order, visit
http://www.haworthpress.com/store/product.asp?sku=5274

or call 1-800-HAWORTH (800-429-6784) in the United States and Canada
or (607) 722-5857 outside the United States and Canada

or contact orders@HaworthPress.com

Reprinted 2009 by CRC Press

Published by

Food Products Press®, an imprint of The Haworth Press, Inc., 10 Alice Street, Binghamton, NY
13904-1580.

Cover design by Lora Wiggins.
Cover photos courtesy of Kim E. Hummer, U.S. Department of Agriculture, Agricultural Research
Service.

**Library of Congress Cataloging-in-Publication Data**

Currants, gooseberries, and jostaberries : a guide for growers, marketers, and researchers in North
America / Danny L. Barney, Kim E. Hummer.
    p. cm.
  Includes bibliographical references and index.
  ISBN 13: 978-1-56022-296-5 (hard : alk. paper)
  ISBN 10: 1-56022-296-4 (hard : alk. paper)
  ISBN 13: 978-1-56022-297-2 (soft : alk. paper)
  ISBN 10: 1-56022-297-2 (soft : alk. paper)
1. Currants—United States. 2. Currants—Canada. 3. Gooseberries—United States. 4. Gooseberries—
Canada. 5. Jostaberry—United States. 6. Jostaberry—Canada. I. Hummer, Kim E. II. Title.

SB386.C9B27 2004
634'.72'097—dc22

                    2004016361

# CONTENTS

# ABOUT THE AUTHORS

**Danny L. Barney, PhD,** received his degree in Pomology from Cornell University in 1987. Professor Barney has been a faculty member of the University of Idaho Plant, Soil & Entomological Sciences Department and Superintendent of the University of Idaho Sandpoint Research & Extension Center since 1988. He is the author or co-author of more than 70 research and university publications, including *Small Fruits for the Home Garden, Ribes Production in North America,* and *Growing Blueberries in the Inland Northwest and Intermountain West.* Professor Barney has conducted varietal trials and research on the physiology and management of currants, gooseberries, and jostaberries since 1989, and has conducted research and written production guides on blueberries, strawberries, raspberries, blackberries, huckleberries, elderberries, and saskatoons.

Professor Barney is a member of the American Society of Horticultural Science and The International *Ribes* Association (TIRA), and has served as Chair of the Northwest Center for Small Fruit Research Genetics technical working group. As a small fruit specialist with Cooperative Extension Service, he has worked closely with commercial fruit growers for more than 15 years. Professor Barney is active in international and national fruit research organizations and presently serves as editor of Web pages devoted to currants, gooseberries, blueberries, and huckleberries on the Northwest Berry and Grape Information Network.

**Kim E. Hummer, PhD,** received her degree in Horticulture from Oregon State University. Since 1987, Dr. Hummer served first as Curator and presently as Research Leader at the U.S. Department of Agriculture's Agricultural Research Service National Clonal Germplasm Repository in Corvallis, Oregon. In this capacity, she is responsible for collecting and maintaining plant materials from around the world. Her broad research assignment includes the study of genetics and germplasm of temperate fruits, nuts, and minor crops, specifically *Actinidia, Asminia, Corylus, Fragaria, Humulus, Mentha, Pyrus, Ribes, Rubus, Vaccinium,* and other related genera. Her research interest concerns the evaluation and characterization of *Ribes* species and cultivars for morphological, phonological, and disease traits.

Since 1996, Dr. Hummer has authored or co-authored more than 40 fruit-related research and educational publications for such venues as *Plant Systematics and Evolution, HortScience, Register of New Fruit and Nut Varieties List 39,* the *Journal of the American Pomological Society, Small Fruits Review, Biotechnology in Agriculture and Forestry,* and the *Fruit Varieties Journal.* She was recently elected as Chair for the Commission on Plant Genetic Resources for the International Society for Horticultural Science. Dr. Hummer has served as an associate editor for *HortScience* and as second Vice-Chair for the American Pomological Society. In 2001, she received a "Super-Supervisor" award from the Association for Persons with Disabilities in Agriculture.

# Foreword

Early in the twentieth century more than 7,000 acres of currants were growing in the United States. Then white pine blister rust was introduced on infected pine seedlings from Europe. The rust caused little damage to currants, but it infected and killed five-needle pine trees, many of which are important lumber species. For that reason, *Ribes* growing was banned in the United States.

The development of currants and pines resistant to rust made it possible to revive the industry, and in 1966 restrictions on currant growing were lifted in most states. Yet in 1989 U.S. currant production amounted to a mere 100 metric tons. At that time, Europe reported a production of 453,000 metric tons.

This book was written because of increased interest in the United States and Canada and the need for up-to-date information on *Ribes* culture and marketing. The authors have been active in evaluating the worldwide *Ribes* collection at the National Clonal Germplasm Repository in Corvallis, Oregon.

The authors have reviewed briefly the history of *Ribes* culture, including currants, gooseberries, and jostaberries (currant-gooseberry hybrids). They present in detail the necessary components of successful culture, including site and soil selection, design of planting, plant propagation, cultivar selection, cultural practices, and pest and disease management, harvesting, and marketing.

Fruit storage requirements are given, and nutritional values of various species are presented. Currants and gooseberries are generally high in vitamin A, and black currant is especially high in vitamin C.

Of special value to plant breeders is a chapter on breeding that lists a wide array of genetic traits for numerous cultivars.

This book is a reliable guide for prospective growers and provides valuable information to *Ribes* researchers.

*M. N. Westwood*
*Professor Emeritus*
*Oregon State University*

# Preface

At the turn of the twentieth century, currants and gooseberries were widely grown in North America, with more than 7,000 acres of fields in commercial production. The U.S. currant production value was more than $1.4 million in 1919 (Hedrick, 1925). Unfortunately, white pine blister rust, an Asian disease that had moved to Europe, was introduced to North America in the late 1800s and early 1900s. This disease spelled doom for North American currant and gooseberry industries. Although the rust disease is not particularly harmful to currants and gooseberries, it can injure or kill five-needled pines. With valuable timber resources threatened, state and federal legislators moved to ban or restrict *Ribes* production throughout the United States. Programs to eradicate native currant and gooseberry populations were established.

Most of the eradication attempts failed because of the wide distribution of native *Ribes* species in the United States and Canada. National restrictions on *Ribes* production were lifted in 1966, although twelve states continue to restrict *Ribes* importation and/or production in the twenty-first century. In light of improved understanding of blister rust ecology and management practices and the availability of rust-resistant and rust-immune currant and gooseberry cultivars, states are reevaluating *Ribes* restrictions. Presently, fruit growers in the United States and Canada are expressing interest in commercial currant and gooseberry production, as well as in production of jostaberries (black currant–gooseberry hybrids). Little up-to-date information on how to grow or market *Ribes* crops is available to growers on this side of the Atlantic. Furthermore, although they are widely popular in parts of Europe, *Ribes* crops are relatively unknown in North America; thus, marketing them to processors and consumers can be difficult. This book provides information for the establishment, maintenance, and marketing of currants, gooseberries, and jostaberries in North America. It also serves as a technical resource for *Ribes* breeders and researchers.

# Chapter 1

# History of Currant, Gooseberry, and Jostaberry Cultivation

## *CURRANTS*

Currants and gooseberries have been used for centuries as food and medicine. Currants were originally known as "corans" or "currans" in England, probably because the fruits resemble "Corinth" grapes (Card, 1907; Hedrick, 1925). Red currants were also known as "red gooseberries," and the English name was "beyond-the-sea gooseberries." The French and Dutch names *groseilles d'outre mer* and *over-zee* indicate that the fruits may have been imported, possibly by the Danes and Normans (Card, 1907), although this point is disputed in the literature. Regardless of their origin, red currants were collected from the wild for medicinal use as early as the 1400s. Sturtevant (1887, cited by Hedrick, 1925) compiled one of the most complete histories of the domestication of currants. Many of the following citations were originally noted by Sturtevant.

Red and white currants were probably first cultivated as garden plants in Holland, Denmark, and the coastal plains surrounding the Baltic (Hedrick, 1925). One of the first herbalists to write about currants was the French author Ruellius in 1536, who believed they made fine border plants and produced fruit suitable as appetizers. Shortly afterward, Ammonius (1539) referred to currants as cherished garden plants. Fuchsi (1542), Tragus (1552), Pinaeus (1561), Camerarius (1586), and Dalechamp (1587) published drawings that resembled 'Common Red', a name that apparently referred to red currant seedlings. By 1558, Matthiolus wrote of currants being common in gardens. Mizaldus (1560) also discussed red currants. In 1576 and 1591, Lobel published drawings of what may have been improved red currant cultivars and mentioned a sweet currant. In his

1586 translation of Dododeus, Lytes (cited by Hedrick, 1925) recognized currants as being in England but translated one name as "beyond-the-sea gooseberry," possibly referring to a foreign origin for the fruit. Camerarius (1587) apparently wrote one of the first currant production guides by providing his readers with directions for sowing currant seeds in gardens. In his guide, Camerarius also mentioned a large-fruited red currant plant growing in the archduke of Austria's garden. Sturtevant (1887) considered the archduke's plant to be the earliest indication of improvement in cultivated red currants and conjectured that this example of *Ribes baccis rubris majoribus* may have been the 'Red Dutch' cultivar or its prototype. In Bauhin's 1858 edition of Matthiolus appeared a drawing of what may have been the 'Red Dutch' cultivar, as well as mention of white-fruited and sweet currants.

The public's unfamiliarity with currants was apparent in Gerarde's *Herball, or General Historie of Plants* (1597, cited by Hendrick, 1925, p. 244). While writing about gooseberries, Gerarde stated, "We have also in our London gardens another sort altogether without prickes, whose fruit is verie small, lesser by much than the common kinde, but of a perfect red color, wherein it differeth from the rest of his kinde." Gerarde was probably describing a red currant. By the mid-1600s, red, white, and black currants were well known in England. Rea (1665, cited by Hedrick, 1925) described five types of currants: small black, small red, great red, red Dutch, and white. He considered black currants as "not worth planting" and small red currants to be "of no better esteem." The red Dutch currants were noted for their large size and sweet flavor. Rea also favored white currants, describing them as "well tasted."

Phillips (1820) described red and white currants as being popular for desserts and considered them cooling and soothing to the stomach. They were admired for their "transparent beauty" and considered valuable for medicinal qualities. According to Phillips (1820, cited by Hendrick, 1925, p. 246), currants were

> moderately refrigerant, antiseptic, attenuant, and aperiant. They may be used with advantage to allay thirst in most febrile complaints, to lessen an increased secretion of bile, and to correct a putrid and scorbutic state of the fluids, especially in sanguine temperments: but in constitutions of a contrary kind, they are apt to occasion flatuency and indigestion.

White currants were also used to make wine in the 1800s. Phillips (1820) considered these wines similar to Grave and Rhenish wines, being best suited as summer table wines. One advantage was their cost, apparently being less expensive than cider. The wines were sometimes diluted for a cooling summer drink and were used to produce a lemonade-like drink called "shrub." Currant juices were used to make acidic punches that were popular in Paris coffeehouses in the late 1700s (Phillips, 1820).

Herbalists interested in their medicinal qualities originally harvested black currants from the wild. In 1611, Tradescant imported them from Holland to the United Kingdom (Brennan, 1996). During the 1700s, black currants were domesticated in eastern Europe and sold at farmers' markets in Russia. According to Hedrick (1925), early botanists generally included black currants in their discussions with red and white cultivars but were more often interested in the medicinal qualities, rather than the fruits' culinary qualities. Because of their medicinal value, black currants were used in most northern European countries. In England, black currants were once called "squinancy berries" because they were used to treat quinsy (tonsillitis). Recent research supports the ancient herbalists' belief that black currants benefit human health. The fruits contain large amounts of vitamin C, anthocyanins, and other antioxidants. The seeds are rich in gamma-linolenic acid (Brennan, 1990).

Despite their medicinal value, black currants were not originally appreciated for their palatability. One herbalist described the fruit as "of a stinking and somewhat loathing savor" (Hedrick, 1925, p. 253). Hedrick seems to have agreed with his ancient colleague, for he wrote: "Few Americans born in the country have tasted the fruit, or ever having done so care for a second taste." Hedrick believed that the farther north black currants were grown, "the less disagreeable the odor and taste, and the larger the currants." This view may well be justified, because the black currant subspecies *Ribes nigrum* var. *sibiricum* from northern Russia presents some of the most valuable germplasm for black currant breeders today. Despite Hedrick's lack of enthusiasm for black currants, they were popular in Canada during the late 1800s and early 1900s and remain popular throughout Europe. Today, black currants are, by far, the leading *Ribes* crop worldwide. Because of their strong flavor, the fruits are nearly always processed rather than being used fresh.

Black currant cultivars have changed markedly since the mid-1900s. Although popular for their flavor and large fruits, western European cultivars, such as 'Baldwin', are often susceptible to diseases and particularly to damage from spring frosts. Although lacking some of the desirable fruit qualities of their western and southern cousins, many Nordic and Russian species and cultivars possess winterhardiness and resistance to spring frost damage. The Finnish cultivar 'Brodtorp' and Swedish cultivars 'Ojebyn' and 'Janslunda' were used in crossbreeding during the development of the immensely successful 'Ben' series at the Scottish Crop Research Institute (Brennan, 1996). Nordic and Russian germplasm have also played important roles in the development of pest and disease resistance. Canadian blister rust–resistant black currant cultivars 'Consort,' 'Coronet,' and 'Crusader' were developed using the Russian species *R. ussuriense.* The newer 'Titania' combines blister rust resistance from both *R. dikuscha* and *R. ussuriense* but has been reported to exhibit rust symptoms in Denmark. Black currant breeding remains a high priority in Europe, with improved pest and disease resistance being primary goals. Perhaps the most serious threat faced by European black currant producers is a virus or viruslike pathogen that produces reversion disease. Many of today's black currant breeding programs include resistance to reversion and its vector, a microscopic mite (Brennan, 1996).

Cultivated currants were introduced into North America in 1629 with the importation of the red currant cultivar Red Dutch (Shoemaker, 1948). A memorandum by the Massachusetts Company, dated March 16, 1629 (Massachusetts Company, 1629), stated that the company was providing for the interests of the colony in the New World:

> To provide to send for New England, Vyne Planters, Stones of all sorts of fruites, as peaches, plums, filberts, cherries, pear, aple, quince kernells, pomegranats, also wheate, rye, barley, oates, woad, saffron, liquorice seed, and madder rootes, potatoes, hop rootes, currant plants.

White and black currants were probably introduced into North America as early as the reds. Although white currants found some favor in the United States, black currants did not, and Hedrick (1925) devoted only a page-and-a-half to the history of its domestication.

In his landmark series *Small Fruits of New York,* Hedrick (1925) described the evolution of currant cultivars in North America. In 1770, the Prince Nurseries in Flushing, Long Island, New York, offered 'Large Red', 'Large White', and 'Large Black' currants in their catalog. By 1806, currant cultivars available in the United States included 'Common Red', 'Large Red', 'Pale White Dutch', 'Large White', and 'White Crystal'. Downing (1845, cited by Hendrick, 1925, p. 249) described ten cultivars, eight of which "are totally undeserving a place in the garden, when those very superior sorts, the White and Red Dutch can be obtained." By 1857, Downing had increased his list of currants to 25, of which he considered the European types to be worthy of brief descriptions. In 1925, Hedrick described 185 cultivars of currants in America, 109 of which originated on this side of the Atlantic.

In 1935, having seen the North American currant industry virtually destroyed by restrictions relating to white pine blister rust, Canadian fruit breeders in Ottawa set out to develop rust-resistant black currants (Hunter, 1950). Using rust-susceptible (but otherwise commercially acceptable) black currant cultivars 'Kerry' and 'Boskoop Giant', breeders incorporated the rust-resistant *Cr* gene from the Siberian species *Ribes ussuriense* Jancz. Two cultivars, 'Crusader' and 'Coronet', were released and widely tested but proved unsatisfactory due to poor self-fertility and were replaced in 1951 by 'Consort' (Kerry × *R. ussuriense*) (Hunter, 1955). Although resistant to rust, 'Consort' proved highly susceptible to American powdery mildew, *Sphaerotheca mors-uvae* and/or *S. macularis,* and produced small berries of inferior quality. Recently, the rust-resistant black currant cultivar 'Titania' was introduced into North America from Europe and appears to be a suitable replacement for 'Consort'. Recommended red, black, and white currant cultivars are described in Chapter 7.

## *GOOSEBERRIES*

Wild gooseberries, from which domestic European cultivars were developed, grow wild throughout temperate Europe and Asia and in the mountains of Greece, Italy, Spain, and northern Africa (Hedrick, 1925). The fact that the wild fruits are small, sour, and could be grown only at high elevations in Mediterranean countries helps explain their being ignored in favor of the abundant grapes and tree fruits available in the same regions. Gooseberries were, reportedly, first domesticated for garden use in Holland but soon became popular in England and France. According to legend, the name gooseberry was derived from the berries being used as a sauce for "green goose." George W. Johnson (cited by Card, 1907, p. 397) disagreed, stating:

> It is somewhat unfortunate for this derivation that it has never been so used. It seems to me more probable to be a corruption of the Dutch name Kruisbes or Gruisbes. Kruisbes, I believe, was derived from Kruis, the Cross, and Bes, as Berry, because the fruit was ready for use just after the Festival of the Invention of the Holy Cross; just as Kruis-haring, in Dutch, is a herring caught after the same festival.

Gooseberries were probably first introduced into England during the reign of Queen Elizabeth (1558-1603), described by Hedrick (1925) as a golden age of gardening in England. In 1548, Turner (cited by Hedrick, 1925) recorded the fact that gooseberries were grown in English gardens. In his *Paradisi in Sole* (1629, quoted by Hedrick, 1925), John Parkinson provided what may be the most complete treatment of early English gooseberries, describing three red, a blue, and a green gooseberry. Regarding their uses, Parkinson (cited by Hendrick, 1925, p. 314) wrote:

> The berries of the ordinary Gooseberries, while they are small, greene and hard, are much used to bee boyled or scalded to make sawce, both for fish and flesh of divers sorts, for the sicke sometimes as well as the sound, as also before they bee neere ripe, to bake into tarts, or otherwise, after manie fashions, as the cunning of the Cooke, or the pleasure of his commanders will appoint. They are a fit dish for women with childe to stay their longings, and to procure an appetite unto meate. The other sorts are not used in Cookery that I know, but serve to bee eaten at pleasure; but in regard they are not so tart before maturity as the former, they are not put to those uses they be.

Credit for improving the gooseberry can largely be attributed to gooseberry enthusiasts in England who competed fiercely in producing prizewinning gooseberries, holding 171 gooseberry shows in 1845 alone (Hedrick, 1925). Large berry size was the primary goal of the gooseberry fanciers, but flavor, beauty, and productiveness were also important (Card, 1907). By 1810, more than 400 gooseberry cultivars had been developed (Harmat et al., 1990).

Gooseberries were introduced into North America at about the same time as currants, but European cultivars proved very susceptible to American powdery mildew, which made growing them in North America difficult. In 1833, the gooseberry picture changed with the discovery of a mildew-resistant seedling by Abel Houghton of Lynn, Massachusetts. The seedling, which was derived from a North American *Ribes* species, was named 'Houghton' and quickly became the mainstay of North American gooseberry growers, even though the berries were small and had only fair quality (Hedrick, 1925; Shoemaker, 1948). The cultivar 'Downing', a seedling of 'Houghton' developed in 1855 by Charles Downing, is considered to be a hybrid between American and European species, and has larger and better fruit than 'Houghton'. Its resistance to mildew and improved fruit quality, along with Downing's ease of propagation, soon made it the leading gooseberry cultivar in North America (Hedrick, 1925). The number of gooseberry cultivars is tremendous. Harmat and colleagues (1990) cited estimates that 4,884 red, yellow, green, and white gooseberries had been named. Identification of specific cultivars is often impossible. Multiple names for a single cultivar are common, and different cultivars may be known by a single name.

## *JOSTABERRIES*

Efforts to develop a hybrid between black currants and gooseberries are not new. A key driving force behind these efforts has been the desire to produce gooseberry-type fruits on thornless plants. A Mr. Culverwell of Yorkshire, England, made one of the first successful crosses between a European black currant and European gooseberry in 1880 (Culverwell, 1883). The hybrid was known as *Ribes culverwellii*. Knight and Keep (1957) cited efforts by later researchers to produce black currant–gooseberry hybrids. Direct crosses between the species produced diploid seedlings that exhibited a wide range of characteristics, with some seedlings resembling gooseberries, others black currants, and yet others being intermediate. Unfortunately, the diploid seedlings were sterile and generally produced little or no fruit, although a few set fruit parthenocarpically (without fertilization) (Duka, 1940; Gorskov, 1940).

German hybridization programs proved more successful than those elsewhere. The efforts began in 1926 with Dr. Paul Lorenz of the Kaiser Wilhelm Institute in Berlin. Lorenz reported that successful crosses were rare and only with the black currant as the seed or mother parent. Within 13 years, approximately 1,000 $F_1$ hybrid seedlings had been established, but the devastation of World War II destroyed all but eight of the seedlings. The eight survivors eventually became part of the Erwin Bauer Institute at Voldsagsen-Hannover, which was established in 1946 (Kennedy, 1990). Dr. Randolph Bauer next took up the challenge of the hybridization program. Using colchicine, Bauer was able to double the number of chromosomes and produce fertile hybrids. Starting with the eight seedlings developed by Lorenz, Bauer backcrossed currant and gooseberry parents to produce a new generation of $F_2$ seedlings. Of the 15,000 fertile seeds produced in these crosses, Bauer eventually selected three superior seedlings, based upon their vigor, fertility, and disease resistance. Two of the seedlings were hybrids of 'Long Bunch' × *R. divaricatum,* while the third was from 'Silvergieters' × 'Green Hansa' (Kennedy, 1990). Bauer called his creations jostaberries, combining the German names *Johannisbeere* (currant) and *stachelbeere* (gooseberry).

The names of the three most common gooseberry–black currant hybrid cultivars, all developed by Bauer, have been confused, and all seem to have marketed under the name 'Josta'. There is, in fact, a cultivar called 'Josta', which was the first to be introduced into the United States. The other cultivars were named 'Jostine' and 'Jogranda' (the latter also known as 'Jostagranda' and 'Jostaki'). These cultivars, introduced into the United States in the early 1990s, were tested in quarantine at the National Plant Germplasm Quarantine Office, Beltsville, Maryland. They are now available through the National Plant Germplasm Repository in Corvallis, Oregon.

Jostaberries exhibit a combination of black currant and gooseberry traits. The stems are spineless and tend to sprawl. The berries are intermediate in size between black currants and gooseberries and, because of chlorophyll in the skins, have dull finishes, lacking the translucence of red and white currants. The flavor is pleasant, but mild, and lacks the strong, characteristic black currant taste.

Plants grown in the United States have, so far, been resistant to American powdery mildew, proving more resistant than any gooseberry or black currant cultivars to the disease in the authors' trials. Jostaberries are also resistant to white pine blister rust, although when sprayed with more than 30,000 spores per milliliter, leaves developed a few uredia (Hummer, unpublished data).

Jostaberries are easy to propagate and very vigorous. The fruits hold on the bushes without overripening for several weeks and freeze well for storage. Jostaberry fruits make good preserves and are suitable for fresh use.

On the negative side, much pruning is needed to control the rank growth. Reports to the authors from around the United States suggest that fruit set can be a problem, even when 'Josta' is planted near a wide selection of black currant and gooseberry cultivars. Yields appear to be lower than for other *Ribes* crops. Enthusiasm for jostaberries in North America has, so far, been limited.

## COMMERCIAL RIBES PRODUCTION

### North America

Currants and gooseberries were widely grown in the United States and Canada during the 1800s and early 1900s. The 1920 U.S. census showed approximately 7,400 acres in commercial production (Hedrick, 1925). Most of the fruits were grown in the middle Atlantic, upper Midwest, and Northeast states, with small farms scattered throughout the rest of the continental United States. According to Shoemaker (1948), New York State was the leading producer of currants, with almost half of the U.S. acreage. Most of New York's production was in the southeastern part of the state, in Ulster and Orange Counties, with additional production along the south shore of Lake Ontario. Berrien County, Michigan, reportedly followed New York in currant production. Currants were also commercially produced in Pennsylvania, Ohio, Minnesota, Missouri, Colorado, Washington, and California. Shoemaker further reported that Marion County, Oregon, was the only county in the United States with more than 100 acres of commercial gooseberries. Other important gooseberry production areas were Polk and Linn counties in Oregon, Oceana and Berrien Counties in Michigan, King and Whatcom Counties in Washington, and Larimer County in Colorado. In Canada, about 50 percent of the currants and gooseberries were grown in Ontario, 25 percent in British Columbia, and the remainder in Quebec and Nova Scotia.

Nearly all of the currant production in the United States was made up of red and white cultivars (Hedrick, 1925). Black currants were never popular in the United States, although they made up approximately 50 percent of the *Ribes* production in Canada, with red currants and gooseberries each making up about 25 percent (Shoemaker, 1948).

Gooseberry production in North America has long been limited by American powdery mildew (Card, 1907; Darrow, 1919; Hedrick, 1925). Partly because of mildew problems, some small fruit experts concluded that gooseberries were better suited to home production rather than commercial farming (Hedrick, 1925; Van Meter, 1928). A common misconception during the 1800s and early 1900s probably contributed to gooseberries' lack of popularity. People believed the berries had to be harvested and used green or they would not jell properly in jams, pastries, and other products. Unripe gooseberries, however, are sour and unpalatable. Card (1907, p. 372) believed that "if people could be accustomed to the use of the ripe fruit, there is no reason why the consumption of the gooseberry should not be immensely increased." Card further reported that markets were seldom overstocked with gooseberries, yet P. T. Quinn stated at the Pennsylvania Fruit Growers' Association meeting in 1872 (cited by Card, 1907, p. 368) that he had seen a thousand barrels of gooseberries "thrown overboard for want of a market." A disgruntled fruit grower added that "they ought to be all dumped into the river."

Sears (1925, p. 268) echoed Card's sentiments about the use of green gooseberries and believed the same logic applied to currants:

> Most housewives prefer that the currants be decidedly unripe, thinking that they will not jell well when ripe. As a matter of fact, they will jell in a perfectly satisfactory manner so long as they are not overripe, and the quality of the jelly will improve very decidedly as the currants pass from the unripe to the perfectly ripe stage. . . . If quality and not appearance is the test,

probably nine people out of ten would vote for the jelly made from ripe currants.

Commercial currant and gooseberry production continued in North America until the late 1920s when white pine blister rust became a major problem for timber industries. The rust was introduced on shipments of white pine nursery stock from Europe to more than eleven U.S. states and three Canadian provinces. Eastern white pine, western white pine, and sugar pine were all valuable timber species threatened by the disease (Darrow, 1946). Common European black currants are the most susceptible *Ribes* host, while domesticated red and white currants and gooseberries are relatively resistant to blister rust. Many *Ribes* species native to North America are susceptible to the disease, which is now firmly established in the wild.

Because white pine blister rust threatened valuable forest resources and gooseberries and currants were minor crops, the U.S. federal government and many state governments took steps to prevent *Ribes* cultivation and even to eradicate native currant and gooseberry species. Although eradication efforts did reduce the incidence of the disease in some areas, most programs were considered ineffective and eventually abandoned (Carlson, 1978). In time, breeders produced pines and black currants that were resistant to blister rust. The federal regulation prohibiting *Ribes* cultivation in the United States was repealed in 1966. Table 1.1 summarizes state restrictions on the importation and/or growing of currants and gooseberries. Be aware, however, that regulations are subject to change, and growers are responsible for complying with them. Canada requires currants or gooseberries from the United States to be accompanied by an import permit and phytosanitary certificate (Dale, 1992). Canada has no national or provincial restrictions on growing currants, gooseberries, or jostaberries. *Ribes* plants and plant parts from Europe or other countries where gooseberries and currants are produced must go through testing at the National Plant Germplasm Quarantine Center in Beltsville, Maryland, prior to entry into the United States. *Ribes* may, reportedly, be imported into the United States from Canada if accompanied by a phytosanitary certificate. At the time of this writing, however, terrorist acts and threats have prompted increased security measures on imports. Prior to importing plant materials into the United States or Canada, growers are advised to contact their respective agricultural departments for the most recent importation requirements.

TABLE 1.1. State regulations governing the importation and growing of currants, gooseberries, and jostaberries, as listed by the U.S. Department of Agriculture.

| State | Regulations |
| --- | --- |
| Alabama | No restrictions listed. |
| Alaska | No restrictions listed. |
| Arizona | No restrictions listed. |
| Arkansas | No restrictions listed. |
| California | No restrictions listed. |
| Colorado | No restrictions listed. |
| Connecticut | No restrictions listed. |
| Delaware | Plants of *Ribes* species from all states are regulated. Shipment into Delaware will be permitted in that portion of the state south of the Chesapeake-Delaware Canal provided a permit authorizing shipment into this area and planting and propagation within the area is obtained from the Delaware Department of Agriculture Plant Industries Section. |
| Florida | No restrictions listed. State regulations may require that cultivated currants or gooseberries be inspected two to three times per year to check for pests and disease. |
| Georgia | No restrictions listed. |
| Hawaii | No restrictions listed. |
| Idaho | No restrictions listed. |
| Illinois | No restrictions listed. |
| Indiana | No restrictions listed. |
| Iowa | No restrictions listed. |
| Kansas | No restrictions listed. |
| Kentucky | No restrictions listed. |
| Louisiana | No restrictions listed. |
| Maine | *Ribes nigrum* is prohibited from all counties. Possession and planting of all other species of *Ribes* is prohibited in the following counties: Androscoggin, Cumberland, Hancock, Kennebec, Knox, Lincoln, Sagadahoc, Waldo, York, and parts of Aroostook, Franklin, Oxford, Penobscot, Piscataquis, Somerset, and Washington. |
| Maryland | No restrictions listed. |

| State | Regulations |
|-------|-------------|
| Massachusetts | *Ribes nigrum* is prohibited from all parts of the state. Possession and planting of all other *Ribes* spp. is prohibited from many towns in Massachusetts. Contact the Massachusetts Department of Food and Agriculture Bureau of Farm Products and Plant Industries for a list of towns. A control area permit issued by the Bureau of Farm Products and Plant Industries must accompany entry of *Ribes* species into unregulated areas. |
| Michigan | Sale of black currant roots, cuttings, or plants is prohibited. Rust-resistant currant varieties may be sold under special permit if approved by Michigan Department of Agriculture. |
| Minnesota | No restrictions listed. |
| Mississippi | No restrictions listed. |
| Missouri | No restrictions listed. |
| Montana | No restrictions listed. |
| Nebraska | No restrictions listed. |
| Nevada | No restrictions listed. |
| New Hampshire | Plants of all species and varieties of *Ribes* are prohibited entry into New Hampshire. |
| New Jersey | *Ribes nigrum* is allowed only under special permit from the New Jersey Department of Agriculture. The movement of all other species of *Ribes* species and *Grossularia* species is prohibited into the following townships: Montague, Sandyston, Walpack, and Vernon in Sussex County; West Milford, Ringwood, and Wanaque in Passaic County; Jefferson in Morris County. |
| New Mexico | No restrictions listed. |
| New York | No restrictions listed. |
| North Carolina | All wild and cultivated currant and gooseberry plants are regulated, and the sale, growing, or planting of currants and gooseberries in North Carolina is prohibited. |
| North Dakota | No restrictions listed. |
| Ohio | All white pine blister rust susceptible varieties of black currant, *Ribes nigrum,* are prohibited. |
| Oklahoma | No restrictions listed. |
| Oregon | No restrictions listed. |
| Pennsylvania | No restrictions listed. |

TABLE 1.1 *(continued)*

| State | Regulations |
| --- | --- |
| Rhode Island | *Ribes nigrum, R. odoratum,* and *R. aureum* are prohibited in all parts of the state. All five-needled pine and other *Ribes* species can be planted only after obtaining a permit from the Rhode Island Department of Environmental Management Agriculture and Resource Marketing. |
| South Carolina | No restrictions listed. |
| South Dakota | No restrictions listed. |
| Tennessee | No restrictions listed. |
| Texas | No restrictions listed. |
| Utah | No restrictions listed. |
| Vermont | No restrictions listed. |
| Virginia | Because *Ribes nigrum* plants are capable of harboring and disseminating the destructive disease of white pine, commonly known as white pine blister rust *(Cronartium ribicola),* European black currant plants *(Ribes nigrum)* may not be moved from any state to any destination in Virginia. |
| Washington | Shipments of fresh currant fruits into certain Washington regions are regulated to control the plum curculio insect. |
| West Virginia | The introduction and dissemination of *Ribes* species is prohibited in the following 23 counties: Barbour, Fayette, Grant, Greenbrier, Hampshire, Hardy, Harrison, Marion, Mercer, Mineral, Monongalia, Monroe, Nicholas, Pendleton, Pocahontas, Preston, Raleigh, Randolph, Summers, Taylor, Tucker, Upshur, and Webster. *Ribes nigrum* is prohibited in all counties of the state. |
| Wisconsin | No restrictions listed. |
| Wyoming | No restrictions listed. |

*Source:* Adapted from the National Plant Board Federal and State Quarantine Summaries (USDA, 2003a).
*Note:* Most states have regulations governing import and sale of nursery stock in general. Only quarantines and regulations that apply specifically to *Ribes* species are included in this table.

The United Nations Food and Agriculture Organization database (FAO, 2002) recorded 170 acres of currants and no gooseberry production in the United States in 1970. By 1980, currant acreage had fallen to 79 acres. No *Ribes* crops were reported by the FAO in North

America in either 1990 or 2000. Although some commercial acreage exists in North America, it is small and scattered. Fruit growers interested in diversifying and adding *Ribes* crops to their farms have frequently contacted the authors of this book. Although the crops have commercial potential, markets in North America are presently limited, and consumer education and marketing are serious factors to address when considering currant, gooseberry, or jostaberry production. For example, during the early 1990s, a major juice company in the United States marketed a cranberry-black currant juice drink. The drink, however, lacked a distinguishable black currant flavor and demand for the product was limited. Eventually the cranberry-currant drink was withdrawn from the market. Home production of currants and gooseberries increased during the 1990s, but the crops still trail blueberries, raspberries, strawberries, and grapes in popularity. Given that black currants are rich in anthocyanins and nutritionally important compounds, great market potential exists for black currant products in North America where consumers are increasingly concerned with slowing the aging process and preventing cancer and heart disease.

## Worldwide

European *Ribes* production continued during the twentieth century, despite industry problems in North America. Tables 1.2 and 1.3 show approximate worldwide currant and gooseberry production, respectively. During the Cold War era, agricultural reports from Soviet block nations were not always reliable. Post-Soviet wars and political reorganizations in Europe have also made reporting difficult.

The overall trend in recent decades appears to have been a steady increase in currant production from 320,358 metric tons in 1970 to 573,025 metric tons in 2000 (FAO, 2002). During the 1990s, currant production fluctuated from about 500,000 to 600,000 metric tons and increased to above 650,000 metric tons in 2001. The large production has caused fruit prices worldwide to fall.

As mentioned earlier, black currants are, by far, the leading *Ribes* crop in the world and are used mostly for juices. In Britain, for example, 75 percent of the crop is juiced. Other popular black currant products rich in anthocyanins (black or purple pigments) and vitamin C include jams, jellies, liqueurs (such as the French Crème de cassis),

TABLE 1.2. Production of currants worldwide from 1970-2000, in metric tons.

| Country | 1970 | 1980 | 1990 | 2000 |
|---|---|---|---|---|
| Australia | 1,015 | 772 | 405 | 1,000 |
| Azerbaijan, Republic of | * | * | * | 1,000 |
| Bulgaria | 2,000 | 394 | 142 | 100 |
| Czechoslovakia | 20,953 | 21,968 | 33,932 | 0 |
| Denmark | 3,143 | 1,316 | 4,000 | 5,000 |
| Estonia | * | * | * | 1,507 |
| France | 6,370 | 5,915 | 7,519 | 8,382 |
| Germany | 162,324 | 158,070 | 146,538 | 140,000 |
| Hungary | 3,839 | 16,107 | 15,157 | 11,848 |
| Latvia | * | * | * | 3,432 |
| Moldova, Republic of | * | * | * | 1,000 |
| Netherlands | 6,726 | 1,900 | 1,500 | 2,200 |
| New Zealand | 210 | 840 | 2,600 | 2,650 |
| Norway | 15,826 | 19,439 | 18,000 | 1,450 |
| Poland | 51,860 | 111,957 | 130,409 | 146,780 |
| Romania | * | * | * | 3,000 |
| Russian Federation | * | * | * | 208,000 |
| Slovakia | * | * | * | 3,919 |
| Sweden | 0 | 0 | 900 | 300 |
| Switzerland | 0 | 0 | 319 | 370 |
| Ukraine | * | * | * | 19,887 |
| United Kingdom | 21,845 | 20,100 | 16,146 | 11,200 |
| United States | 247 | 40 | 0 | 0 |
| USSR | 24,000 | 80,000 | 70,000 | 0 |
| Total | 320,358 | 438,818 | 447,567 | 573,025 |

*Source:* Based on data from FAO (2002).
*Production figures for 1970-2000 included in USSR data.

and colorings for yogurt and other dairy products (Brennan, 1996). Much of the crop grown in the United Kingdom is used to produce the sweet, black currant nectarlike drink Ribena which was developed during World War II and is increasingly popular today. The Russian

TABLE 1.3. Production of gooseberries worldwide from 1970 to 2000, in metric tons.

| Country | 1970 | 1980 | 1990 | 2000 |
|---|---|---|---|---|
| Czechoslovakia | 7,730 | 9,600 | 13,802 | 0 |
| Denmark | 229 | 100 | 0 | 0 |
| Germany | 94,516 | 80,979 | 84,182 | 75,000 |
| Hungary | 6,324 | 13,563 | 8,220 | 4,649 |
| Moldova, Republic of | * | * | * | 1,200 |
| Netherlands | 352 | 160 | 0 | 0 |
| New Zealand | 170 | 40 | 10 | 10 |
| Norway | 3,379 | 3,130 | 3,100 | 170 |
| Poland | 17,600 | 37,023 | 34,848 | 28,583 |
| Russian Federation | * | * | * | 35,000 |
| Slovakia | * | * | * | 1,227 |
| Switzerland | 0 | 0 | 54 | 46 |
| United Kingdom | 12,802 | 7,400 | 2,733 | 1,500 |
| USSR | 35,000 | 67,000 | 50,000 | 0 |
| Total | 178,102 | 218,995 | 196,949 | 147,385 |

*Source:* Based on data from FAO (2002).
*Figures for 1970-1990 included in USSR production data.

Federation, Poland, and Germany lead the world in production of black currants (FAO, 2002). The main red currant producers are Poland and Germany, with lesser production in Belgium, France, Holland, and Hungary. (Brennan, 1996). Germany and the Slovak Republic are the leading producers of white currants. Red currants are used mostly for juices and other processed food items, often in combination with other fruits. White currants are used in parts of Europe for baby food (Harmat et al., 1990) and in Finland for sparkling wines.

Gooseberry producers have fared less well in recent years. Although gooseberry production increased between 1970 and 1980, it declined by 1990 and fell to 147,385 metric tons in 2000 (FAO, 2002). Germany, the Russian Federation, and Poland are the world's largest gooseberry producers. Gooseberry production in Britain has fallen to very low levels because of decreased demand and increased production costs, particularly at harvest time (Brennan, 1990). Gooseberries are used mostly for jams and canned products, but small amounts are also sold for the dessert trade.

# Chapter 2

# Genetics, Growth, Development, and Fruit Composition

## TAXONOMY AND GENETICS

Woody shrubs in the genus *Ribes* are found throughout the northern portion of the northern hemisphere, along the Andes Mountains in South America, and in Mediterranean northwest Africa (Brennan, 1990). Of approximately 150 *Ribes* species worldwide, about 18 have been used in developing modern cultivars (Harmat et al., 1990). In addition to several species that have ornamental landscape use, five subgenera of *Ribes* are grown for their fruit: black currants, red currants, white currants, gooseberries, and jostaberries.

The taxonomy, or naming, of currants and gooseberries has been confused for centuries and remains so today (Spongberg, 1972). A thorough treatment of the genetics and taxonomy of *Ribes* is beyond the scope of this book, which is intended as a practical guide to currant and gooseberry research and production. Readers interested in these subjects, particularly those involved in *Ribes* breeding programs, would do well to refer to "Currants and Gooseberries" (Keep, 1975) and "Currants and Gooseberries *(Ribes)*" (Brennan, 1996). Breeders specifically interested in black currants may also be interested in the article "Black Currant Breeding Using Related Species As Donors" (Keep et al., 1977).

Gooseberries and currants were originally placed in the Saxifragaceae family in 1891 (Engler and Prantl, 1891) but have since been relocated into Grossulariaceae because of their wholly inferior ovaries, totally syncarpous gynoecium, and fleshy fruits (Cronquist, 1981; Lamarck and De Candolle, 1805; Sinnott, 1985). Linnaeus (1737, 1738) originally assigned all currants and gooseberries to the single genus *Ribes,* although some later workers believed that currants should be placed into the genus *Ribes* and gooseberries into *Grossularia* (Berger, 1924; Coville and Britton, 1908; Komarov, 1971). Numerous infrageneric classifications were proposed for these two genera.

Anatomical studies (Stern et al., 1970) provided no justification for two genera, and prevalent monographs recognize a single genus: *Ribes* (de Janczewski, 1907; Sinnott, 1985; Brennan, 1990). Crossability between gooseberry and currant species supports the concept of a single genus (Keep, 1962). Even assuming the single genus *Ribes,* however, the taxonomic picture remains clouded. Some taxonomists have suggested dividing *Ribes* species into subgenera and sections or series. de Janczewski (1907) subdivided the genus into six subgenera: *Coreosma,* the black currants; *Ribes* (= *Ribesia*), the red currants; *Grossularia,* the gooseberries; *Grossularioides,* the spiny currants; *Parilla,* the Andean currants; and *Berisia,* the European alpine currants. Keep (1962) developed a useful classification system for *Ribes* breeders by classifying species into nine subgenera based upon their cross-fertility. Keep's work is described in Chapter 12.

Despite their great diversity in plant characteristics and crossability, all *Ribes* species are diploid, having a basic chromosome number of x = 8 and n = 16. With the exception of a small number of artificially created hybrids in which chromosome numbers have been increased, all currant and gooseberry cultivars are also diploid. Throughout the genus, the chromosomes are small and quite uniform (Sinnott, 1985).

The centers of diversity for the subgenera *Eucoreosma, Ribesia,* and *Berisia* include northern Europe, Scandinavia, and the Russian Federation (Jennings et al., 1987). In addition, several species of black currants with sessile yellow glands are native to South America. Gooseberries *(Grossularia)* have centers of diversity in Europe and North America, depending on species.

*Eucoreosma,* the subgenus for black-fruited currants, has sessile resinous glands. The species of most economic importance is *R. nigrum* L., which is native through northern Europe and central and northern Asia to the Himalayas and includes the subspecies *europaeum* and *scandinavicum,* and *R. nigrum* var. *sibiricum* Wolf (Brennan, 1996). *Ribes nigrum* is an unarmed, strongly aromatic shrub, growing as tall as six feet (2 m) (Rehder, 1986). The leaves are lobed, up to four inches (10 cm) per side, glabrous above, slightly pubescent with numerous sessile, aromatic glands beneath; the racemes droop and have four to ten flowers. The flowers have reddish- or brownish-green campanulate hypanthia and recurved sepals. The whitish petals are about two-thirds as long as the sepals. The fruits are globose, up to 10 mm diameter, and are generally shiny black when ripe, although green- and yellow-fruited forms exist. Recent breeding efforts have doubled fruit size compared to wild fruits.

Breeders today are crossing *R. nigrum* with other species to improve disease resistance and yield. Several black currant cultivars trace their white pine blister rust resistance to *R. ussuriense* Jancz., which is native from Manchuria to Korea. Other species being used in breeding programs to improve disease resistance include *R. dikuscha* Fisch. from eastern Siberia and Manchuria, and *R. nigrum* var. *sibiricum* from northern Europe. Incorporating *Ribes bracteosum* Dougl., the California black currant, into breeding programs has increased cluster length and yields of black currants (Brennan, 1996). *Ribes hudsonianum* Rich., the northern North American black currant, and *R. americanum* Mill., the American black currant, have desirable traits that may be useful for broadening the cultivated black currant gene pool. In North America, the cultivar Crandall is a selection of golden currant, *R. aureum* Pursh var. *villosum* (formerly *R. odoratum*). 'Crandall' is distinctly different from European black currants, having larger and sweeter berries that lack the characteristic black currant flavor and ripen a month or more later than most European cultivars.

Red currants (subgenus *Ribesia*) are characterized by having crystalline glands on young growth (Brennan, 1996). Several red currant species are economically important. Selections of *R. sativum* Syme (= *R. vulgare* Jancz.) were initially made from native stands in northwestern Europe. *Ribes petraeum* Wulf., a mountain-dwelling species, also contributed selections from the wild. Most cultivated red currants are derived from *R. rubrum* L., a Scandinavian species that is native as far north as 70° north latitude (Brennan, 1996). *Ribes rubrum* is an unarmed shrub that grows to six feet (2 m) tall (Rehder, 1986). The shoots are glabrous or have glandular hairs. Stems are covered with a smooth pale yellow bark. The leaves are deeply cordate, three to five lobed, two to three inches (6 to 7 cm) in diameter. Flowers, which occur on long racemes, are greenish tinged with purple. The hypanthium is almost flat and the petals are very small.

The fruit is globose, one-fourth to three-eighths inch (6 to 10 mm) in diameter, red, and glabrous. *Ribes spicatum* Robs., from Norway, and *R. multiflorum* Kit., from England, are used in red currant genetic improvement programs. *Ribes triste* Pallas (North American red currant) has fruit quality similar to European red currants but has not been developed for cultivation. White and pink currants are color variations of the red currant species.

Gooseberry species (subgenus *Grossularia*) have nodal spines that are often called thorns. The European gooseberry, *R. grossularia* L. (synonymous with *R. uva-crispa* L.), is native to the United Kingdom and eastward through northern Europe, the Caucasus, and North Africa. *Ribes grossularia* was the source of early gooseberry cultivars and is frequently selected for cultivar development (Brennan, 1996). This species produces spiny canes that grow as much as five feet (1.5 m) tall and bear two to three spines at the nodes. Leaves are as large as 2.5 inches (6.5 cm) in diameter, sparsely pubescent, or glabrous. Flowers develop in axillary clusters of one to three (much fewer than those on currant racemes), are pale green, sometimes pinkish, and have a hemispherical hypanthium, reflexed sepals, and short white petals. The berries, which can be covered with bristly hairs (hispid), are round to oval in shape, about three-eighths inch (10 mm) in diameter, and green, yellow, or purplish-red. Cultivars have much larger berries than are typically found in the species. American gooseberry species such as *R. divaricatum* Dougl., *R. hirtellum* Michx, and *R. oxyacanthoides* L. are used extensively in gooseberry breeding to improve disease resistance and decrease spininess of the large-fruited European cultivars.

Jostaberries (*Ribes × nidigrolaria* Bauer) are hybrids between gooseberries and black currants. These hybrids are very vigorous, do not have the acrid odor of black currants, have few or no spines, and are disease resistant. Their disease resistance and large size fruit make them popular with organic farmers and home gardeners.

Ornamental and flowering *Ribes* species have a broad range of colors. The American species *R. aureum* Pursh. has fragrant yellow flowers with tubular hypanthia that bloom in spring (Rehder, 1986). The fruits are black but do not have the "black currant odor" characteristic of *R. nigrum*. The black currant cultivar 'Crandall' was selected from *R. aureum*. Another American species, *R. sanguineum* Pursh, has a range of flower color variants, from white to dark red-purple, and is used in landscape plantings for spring bloom and wildlife habitat. Unfortunately, this species tends to be susceptible to white pine blister rust. Some American gooseberry species, such as *R. speciosum* Pursh., *R. lobbii* Gray, and *R. menziesii* Pursh have very attractive fuchsia-like flowers and are planted for their ornamental landscape attributes (Brennan, 1996).

Many currant and gooseberry species and subspecies exist, in addition to those described previously. Species that have been used in domesticating *Ribes,* or which may be of interest to breeders, are listed in Box 2.1. Additional species that might prove useful in breeding new cultivars are listed in Chapter 12.

## GROWTH AND DEVELOPMENT

Cultivated gooseberries and currants are perennial, woody shrubs that normally grow between three and five feet (1 to 1.2 m) tall. Of the

---

### BOX 2.1. *Ribes* Species Used to Develop Existing Currant and Gooseberry Cultivars

**Gooseberries**

| | |
|---|---|
| R. aciculare | R. grossularia |
| R. burejense | (R. uva-crispa) |
| R. cynobasti | R. hirtellum |
| R. divaricatum | R. oxyacanthoides |

**Black Currants**

| | |
|---|---|
| R. americanum | R. nigrum |
| R. aureum | europeanum |
| (R. odoratum) | scandinavicum |
| R. bracteosum | sibiricum |
| R. dikuscha | R. petiolare |
| R. hudonianum | R. ussuriense |

**Red and White Currants**

| | |
|---|---|
| R. longeracemosum | R. sativum (R. vulgare) |
| R. petraeum | R. spicatum |
| R. rubrum | |

cultivated *Ribes,* black currant bushes are generally larger than those of red and white currants, while gooseberry bushes are usually smaller than currants. European gooseberry bushes are smaller and more upright than American types and are more susceptible to powdery mildew problems in North America. Jostaberries are the largest cultivated *Ribes* and can grow to eight feet (2.6 m) or more in height. These are the sizes typically seen in commercial cultivation. When given special care, gooseberries have achieved remarkable proportions. Card (1907, p. 366) reported that a gooseberry trained to a tree form grew 16 feet (5 m) high, while a bush trained to the side of a building "measured 53 feet 4 inches [16 m] from one extremity to the other" and bore 30 to 40 quarts (28 to 38 L) of fruit annually.

Life expectancy of a commercial planting depends largely on the care it receives. Provided that the bushes are fertilized and pruned properly, and pests and diseases are controlled, red and white currants and gooseberries can remain productive for 20 years or more, while black currants have a productive life expectancy of ten or more cropping years (Harmat et al., 1990). Currants and gooseberries tend to remain productive longer in northern regions than near their southern limits (Shoemaker, 1948). Van Meter (1928) noted that gooseberries range a little farther south than currants but are still distinctly cool-climate fruits. Again, these observations apply to commercial production. In North America, it is not unusual to find currants and gooseberries that have survived a century or more on abandoned farmsteads. The London Horticultural Society (cited by Card, 1907, p. 366) reported a gooseberry bush that was 46 years old, with a circumference of 36 feet (11 m) and which "produced several pecks of fruit annually for 30 years."

### Spring Growth

Currants, gooseberries, and jostaberries are among the first plants to start growing in the spring. They are so well adapted to growing in cool spring weather that potted plants can break bud and start growing when stored in the dark at around 35°F (1.7°C). The tendency of *Ribes* to break dormancy and start growing in storage makes fall digging and winter storage difficult. For that reason, most commercial nurseries that ship bareroot stock prefer to dig and ship in the fall for fall planting. Early blooming also makes the flowers of cultivated *Ribes* susceptible to spring frost damage, which contributes to the highly variable yields obtained by European farmers, especially for black currants.

Swelling buds are the first sign of spring growth. The buds begin forming in the leaf axils on new shoots during the summer. By late fall or early winter the buds are fully mature, but they normally do not break and begin growth until they have been exposed to temperatures slightly above freezing for a period of time. The requirement for exposure to chilling temperatures is common in temperate-zone shrubs because it helps ensure that the plants remain dormant throughout winter warm spells and subsequent cold snaps. The amount of chilling the plants need differs among cultivars, but currants and gooseberries typically require 800 to 1600 hours at temperatures between 32 and 45°F (0 and 7°C) (Westwood, 1978). Jostaberries appear to have similar requirements. The requirement for chilling to break dormancy is one reason that currants and gooseberries are generally not well suited to cultivation in southern states.

## Cane Formation

Once the chilling requirement is met and temperatures are favorable, the buds begin swelling and eventually break. There are two kinds of buds, vegetative and flower. Bud formation is discussed in detail later in this chapter. For now, suffice it to say that new shoots develop from vegetative buds. In some cases, the shoots remain very short and set terminal buds. These short shoots are called spurs and are important in red and white currants, gooseberries, and jostaberries because their flowers are largely set on spurs. The canes and shoots of currants and jostaberries are smooth, but many gooseberry cultivars have thin, sharp spines located at the nodes. Spininess differs among gooseberry cultivars, and some have nearly smooth canes.

On the tip of each dormant cane is a terminal or apical bud. These buds are nearly always vegetative and produce new shoots that add length to existing canes. Lateral buds are formed along the stem below the shoot tip and produce side branches. One characteristic of terminal buds is that they produce chemicals called growth regulators, which are similar to hormones in animals. Some of the growth regulators produced in terminal buds prevent the lateral buds on the stem below from breaking and developing new shoots. As a result of this "apical dominance," currant canes tend to remain long and straight with few branches near the tips. Because smaller amounts of growth regulators reach them, lateral buds farther down the stems are less subject to apical dominance than higher buds and are more likely to break and produce branches. If disease, browsing, or improper pruning damages a terminal bud, apical dominance is eliminated and several branches usually begin developing near the tip of the cane. Heavily branched bushes are dense, restrict air movement through the bush, and increase pest and disease problems. Heavily branched bushes are also more difficult to harvest than those with straight canes.

Gooseberries develop similarly to currants, but apical dominance is not as strong. As a result, gooseberry bushes tend to be shorter and more heavily branched than currants. Some gooseberry cultivars produce thin, droopy canes, resulting in low, rounded bushes rather than the vase-shaped bushes that currants often form. European gooseberry bushes are typically more compact than American gooseberries.

The length of new shoots varies greatly, depending upon the cultivar and vigor of the plant. Other factors include weather and moisture conditions, soil fertility, weed control, and disease control. Small gooseberry cultivars may develop shoots six inches (15 cm) long or less, while vigorous black currant and jostaberry shoots may be three feet (1 m) long.

In addition to the shoots that develop from buds on the canes, new canes also develop from the collar (slightly below to slightly above the soil surface) each spring. These new canes are very important because they provide new fruiting wood. In most cultivation systems, canes are pruned off when they are three to four years old. Yields decline as canes age, and large, top-heavy bushes are hard to manage. Regularly replacing older canes helps keep bushes young and vigorous. Depending upon the cultivar, vigor, and training system, from three to six new canes are retained each spring, while the same numbers of older canes are pruned out. Pruning is discussed in detail in Chapter 9.

## Leaves and Buds

As the shoots begin growing in the spring, leaves develop in an alternate arrangement on them. *Ribes* leaves have a characteristic lobed shape that is easy to identify. Canes and leaves are illustrated in Figure 2.1. Leaves produce biochemicals that plants need to grow and develop. Through the process of photosynthesis and subsequent metabolic pathways, plants combine carbon dioxide from the air, water and mineral nutrients from the soil, and energy from sunlight to create sugars and other biochemicals. Leaves damaged by disease, pests, excessive heat, wind, or drought are less able to carry out these metabolic processes than healthy leaves, weakening the plant.

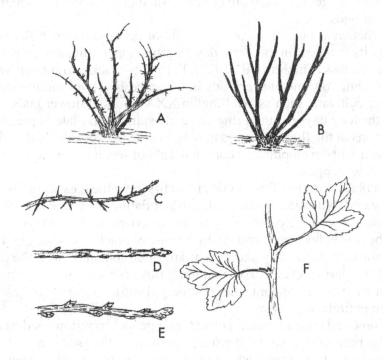

FIGURE 2.1. *Ribes* cane and leaf anatomy: (A) gooseberry bush; (B) currant bush; (C) gooseberry cane; (D) black currant cane; (E) red or white currant cane; (F) typical currant and gooseberry leaves.

*Ribes* leaves are rich in phenolic and terpenoid chemical compounds. In his work with black currants, Marriott (1988) identified 44 chemicals in the leaf oil alone. Black currant leaves and buds are used as medicinal herbs (Harmat et al., 1990). Leaves also produce growth regulators that keep the buds on current season shoots from breaking and beginning growth, much the same way that apical buds reduce branching on one-year-old canes (Tinklin and Schwabe, 1970). If the leaves are removed or severely damaged, the buds will break to form new shoots and leaves to replace the ones that have been damaged.

Within the axil of each leaf (where the leaf connects to the stem) a bud develops. *Ribes* buds are simple, which means that they produce either shoots and leaves (vegetative) or flowers (reproductive), but not both. Vegetative buds are typically smaller and more pointed than flower buds.

Whether a bud develops into a shoot or inflorescence depends upon its location on the stem, day length, temperature, and probably other factors (Tinklin et al., 1970). For many black currant cultivars, differentiation into flower buds is triggered by short autumn days. Other cultivars, such as 'Wellington XXX', initiate flower buds during the long days of late spring and early summer. By late September, portions of the flower clusters may be visible between the bud scales. Flower bud development is completed about seven to ten days before the flowers open.

In black currants, flower buds form first near the base of one-year-old wood, with later flower buds differentiating toward the tip of the shoot. As with the other currants, gooseberries, and jostaberries, all of the flower buds are terminal. In the case of black currants, the flowers form on secondary shoots that are often too small to be readily visible (Harmat et al., 1990) and more flowers form on one-year-old than on two-year-old branches (Heiberg, 1986). Short shoots may remain entirely vegetative.

Some red currant, white currant, and gooseberry flowers develop at the base of one-year-old wood, but most of the fruit comes from terminal buds on spurs, which are located on two- to three-year-old wood. The buds themselves actually form on the one-year-old portion of the spurs. Flower bud formation in jostaberries resembles that in gooseberries more than it does black currants, which is why pruning methods are similar for gooseberries and jostaberries. The difference in flower location is the reason that black currants are pruned differently from other *Ribes*. Pruning is discussed in detail in Chapter 9.

### Flowering, Pollination, Self-Fertility, and Fruiting

Flowers on red and white currants are clustered on long stems or racemes, often called "strigs." Black currant flowers also develop in racemes, but the racemes are typically shorter than in red and white currants. Shorter fruit clusters are one reason black currants yield less than red or white currants. Gooseberry and jostaberry flowers usually develop in small clusters or sometimes singly. Flower clusters are shown in Figure 2.2.

*Ribes* flowers are greenish-yellow or red and are pollinated by bees and other insects. Since pollen is not readily shed during the night or early morning, most of the pollination occurs during the day. According to Baldini and Pisani (1961), most pollen is released from the an-

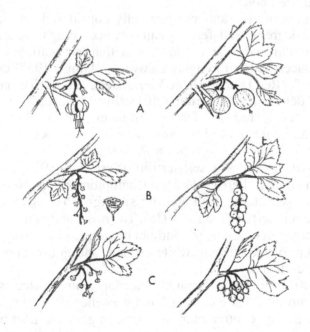

FIGURE 2.2. *Ribes* flower and fruit anatomy: (A) gooseberry; (B) red or white currant; (C) black currant.

thers between 2:00 and 6:00 in the afternoon. The importance of effective pollination cannot be overstated. Szklanowska and Dabska (1993) investigated the effects of excluding bees and other pollinating insects from black currant plants. They found with three black currant cultivars that lack of insect pollination reduced yields by 70 to 80 percent. Field designs and recommendations for beehives are discussed in Chapters 5 and 9.

According to Brennan (1996), self-sterility, also called self-incompatibility, predominates among *Ribes* species in the wild. In other words, fertilizing the flowers of one plant with pollen from the same plant will seldom result in fertilization, and few or no fruits will be produced. Because many domestic currant and gooseberry cultivars represent combinations of species, the rule of self-fertility is less clear than with the species themselves. Although some authorities consider currants and gooseberries to be self-fruitful and the use of pollinizing cultivars unnecessary (Dale and Schooley, 1999), exceptions have been noted.

Red and white currants are generally considered self-fertile, although the degree of self-fertility can vary according to year and location. Seljahudin and Brozik (1967) found that self-pollination significantly reduced fruit set in many cultivars. Klambt (1958) concluded that 'Heros', 'Red Dutch', 'Rote Vierlander', and 'Heinemanns Rote Spatlese' demonstrated partial self-sterility. Gooseberries are also generally considered self-fertile (Brennan, 1996), although self-sterility can be introduced by crossing with wild species. American species *R. robustum* (*R. niveum* × *R. inerme*), *R. succirubrum,* and *R. divaricatum* have low self-fertility (Sergeeva, 1979, 1980). The American cultivar 'Perry' ('Oregon Champion' × *R. missouriense*) is completely self-sterile, although its siblings 'Pixwell' and 'Abundance' are self-fertile (Yeager, 1938). To ensure and enhance fruit set in commercial operations, the authors recommend planting at least 1 to 2 percent of red currant, white currant, and gooseberry fields to pollinizer cultivars.

Self-fertility in black currants is a more complicated issue than with red currants, white currants, and gooseberries, and research results have proven contradictory at times. In general, most black currant cultivars are partially or completely self-fertile. In only two cases have domestic black currants been found to be completely self-sterile: a rogue of 'Invincible Giant Prolific' (Luckwill, 1948) and a form of

'Lee's Prolific' (Ledeboer and Rietsema, 1940). Wellington and colleagues (1921) considered the four main western European black currant groups ('Boskoop Giant', 'French Black', 'Baldwin', and 'Goliath') to be self- and cross-fertile. Fernqvist (1961) determined that 'Goliath', 'Silvergieters Zwarte', 'Wellington X', and 'Wellington XXX' set nearly the same crops with self- or cross-pollination. Free (1968) concluded in British trials that natural cross-pollination in the field did not increase yields for 'Baldwin', 'Cotswold Cross', 'Davison's Eight', 'Mendip Cross', 'Seabrook's Black', 'Wellington XXX', or 'Westwick Choice'.

In Italy, however, Baldini and Pisani (1961) found that while 'Amos Black', 'Goliath', 'Silvergieters Zwarte', and 'Westwick Choice' were highly self-fertile, 'Baldwin Hilltop', 'Boskoop Giant', and 'Wellington XXX' were only partially self-fertile and yields could be improved with cross-pollination. Tamas and Porpaczy (1967) concluded that 'Mendip Cross' and 'Wellington XXX' had relatively low levels of self-fertility. Other researchers have also reported increased black currant yields associated with cross-pollination (Voluznev, 1948; Neumann, 1955; Hofman, 1963; Lucka et al., 1972; Voluznev and Raincikova, 1974).

Tamas (1964) considered northern European black currants to be more self-fertile than western European types. Kuminov (1962), on the other hand, concluded that cultivars derived from *R. nigrum sibiricum* tend to be self-sterile, and Potapenko (1966) believed that cultivars derived from the Siberian wild currant, *R. cyathiforme,* also tend to be self-sterile. 'Noir de Bourgogne' is also highly self-sterile (Lantin, 1970), as are 'Barhatnaja', 'Belorusskaja Pozdnjaja', 'Losickaja', 'Mecta', and 'Minskaja' (Voluznev, 1968). The Canadian cultivars 'Consort', 'Crusader', and 'Coronet' ('Kerry' × *R. ussuriense*) are self-sterile and generally unable to pollinate one another (Keep, 1975). The pollen from 'Crusader' tends to clump together and is not easily picked up by bees (Keep, 1975), while 'Coronet' flowers are narrow at the top and bees often reach in from the side without contacting the anthers.

In summary, most black currant cultivars are partially or completely self-fertile, although there are differences between cultivars and self-fertility may vary according to growing region and, possibly, other factors. Unfortunately, self-fertility or self-sterility has been determined for relatively few black currant cultivars (Box 2.2). For

## BOX 2.2. Self-Fertility or Self-Sterility in Some Common Black Currant Cultivars

### Self-Fertile

Altajskaja Desertnaja[j]
Amos Black [b,i,m]
Baldwin[e,i,o]
Beloruskaja Sladkaja[p]
Black Reward[h]
Boskoop Giant[o]
Brodtorp[l,m]
Cernaja Grozd[j]
Cernaja Lisavenko[j]
Cotswold Cross[e]
Druznaja[j,k]
Dymka[k]
French Black[o]
Goliath[b,d,o]
Golubka[f,j]
Gonoaltaisk[c]
Imandra[c]
Kantata 50[p]
Koska[c,j,k]
Mendip Cross[e]

Minaj Smirev[p]
Moskovskaya Rannjaja[j]
Nina[j]
Paulinka[p]
Pilot A. Mamkin[p]
Pobyeda[j]
Primorskij Cempion[c,g,j]
Seabrooks Black[e]
Silvergieters Zwarte[b,d,l]
Stakhanovka Altaja[j]
Stella I[i]
Tinker[i]
Tor Cross[i]
Uspekh[j]
Vistavochnaja[c]
Wellington X[d]
Wellington XXX[d,e]
Westwick Choice[b,e]
Zoja[j,k]

### Partially Self-Fertile

Baldwin Hilltop[b]
Boskoop Giant[b]
Mendip Cross[b,m]

Rodknop[b]
Wellington XXX[b,m]

### Self-Sterile

Barhatnaja[n]
Bellorusskaja Pozdnjaja[n]
Bogatyr[a]
Consort[f]
Coronet[f]
Crusader[f]

Losickaja[n]
Mecta[n]
Minskaja[n]
Noir de Bourgogne[g]
Wasil[a]

*Source:* [a]Badescu and Badescu (1976), [b]Baldini and Pisani (1961), [c]Elsakova (1972), [d]Fernqvist (1961), [e]Free (1968), [f]Keep (1975), [g]Kuminov (1962), [h]Lantin (1970), [i]Nes (1976), [j]Osipov (1968), [k]Potapenko (1966), [l]Tamas (1964), [m]Tamas and Porpaczy (1967), [n]Voluznev (1968), [o]Wellington et al. (1921), [p]Zazulina (1976).

black currants, the authors recommend planting two or more cultivars together to ensure cross-pollination. Kuminov (1962) considered 'Sinjaja', 'Druznaja', and 'Nocka' to be good pollinizers for self-sterile Siberian cultivars. Unfortunately, the presence of pollinizers does not guarantee cross-pollination. Klambt (1958) noted that open-pollinated black currants were usually self-pollinated, unlike red currants. Like their black currant parents, jostaberries also benefit from cross-pollination.

For effective cross-pollination in all *Ribes* crops, ensure that the main crop and pollinizers are compatible and bloom simultaneously. Because cultivars within each crop bloom over relatively long times and the blooms usually overlap within crops, timing of crop and pollinizer cultivars is seldom a problem. Likewise, compatibility between crop and pollinizer cultivars is generally not a problem, although it is advisable not to use siblings as pollinizers. Details on setting out fields for cross-pollination are given in Chapter 5.

Not all of the flowers on a strig develop at the same time. The flowers near the base of the strig typically develop first, followed by the flowers near the tip. On cultivars with long strigs, the basal flowers may develop and be ready for pollination as many as 20 days earlier than the terminal flowers (Harmat et al., 1990). The difference in the time of pollination means that not all berries ripen at the same time. In black currants structural differences between the flowers at the bases of the strigs and those near the tips result in poorer pollination of the latter. These structural differences are one cause of fruit abortion from the tips of the clusters, which is especially prevalent in 'Boskoop Giant'.

*Ribes* flowers are "inferior," which means that the ovary and developing fruit are found below the point where the flower attaches to the stem. Soon after pollination, fruit begins to develop. According to Westwood (1978), cell division within the fleshy part of the fruit is completed by about the time of full bloom. Once the cells are formed, they begin to expand, and the developing fruit becomes visible as swellings at the bases of the flowers. Immature fruits are hard, green, and sour, but they quickly mature. Black currants ripen after strawberries and with or before red raspberries during the early summer. Depending on location, types of crop, and cultivars, *Ribes* in North America ripen from May through August, with ripening extending over about a four-week period. Ripening may be delayed in colder regions. Weather also plays a factor, and ripening can be delayed by cool weather in any location.

Fruit size is affected by pollination in many crops, and the same can be true for *Ribes*. In an experiment with black currants, Webb (1976) found that berry size was partially determined by the number of viable seeds per berry. This discovery reemphasizes the need for effective pollination. The advisability of planting pollinizer cultivars for all *Ribes* crops was discussed earlier. Planting designs to enhance cross-pollination are described in Chapter 5. Growers can also improve pollination by placing beehives in their fields during bloom and mowing cover crops to remove flowers that might draw pollinators away from the currants and gooseberries. The use of hives and cover crops is discussed in Chapter 9.

In apples and some other tree fruits, growers commonly remove part of the fruit to force the trees' resources into fewer fruit. The retained fruits are larger and more attractive than those from unthinned trees. An experiment conducted with red currants in New Jersey in 1888 showed that cutting off the bottom half of the flower clusters increased fruit set on the remaining portion by 15 percent and fruit size by 7 percent (Halstead, 1890, cited by Bailey, 1897). Gooseberry thinning has been practiced for centuries. In nineteenth-century England, during the heyday of gooseberry competitions, growers considered thinning an "absolute necessity" in producing prizewinning berries (Card, 1907). They typically removed all but a few fruits from the bushes to force the plants' resources into a few "sinks" and develop extremely large berries. Even today, it is common to harvest gooseberries in three pickings. According to advocates of this method, the small, green fruits from the first two harvests are used for processing. The final third of the berries are left on well-exposed branches and develop into large, dessert-quality fruit. Webb (1976) and Fernqvist (1961), however, found that the numbers of flowers or fruits did not affect black currant berry weights, respectively, per strig. As a practical matter, it is not commercially practical to thin currants to increase berry size. Proper nutrition, pruning, irrigation, and other care should suffice to produce large, high-quality fruit. For small plantings, growers may choose to thin gooseberries by picking them in three stages, as described previously. For the fancier, thinning gooseberries is valuable in producing large fruits for competition and display.

Currant berries are smooth and nearly round. Red and white currants have glossy, translucent skins that allow one to see the seeds within the fruit. Black currant, gooseberry, and jostaberry fruits appear dull because they have chlorophyll in their skins. Jostaberries tend to be egg shaped and, like their currant parents, are smooth. Although most popular gooseberry cultivars are nearly or completely smooth, some cultivars produce fruits that are covered with fine, stiff hairs. Gooseberries come in a remarkable array of shapes, sizes, colors, and surface textures that lend themselves to culinary art and visually attractive desserts. Ripe black currant, gooseberry, and jostaberry fruits retain their flower parts, giving the fruits an elongated, bearded look. Red and white currant berries often do not retain the flowers. *Ribes* fruits are illustrated in Figure 2.2.

Red and white currant fruits have tender skins that tear when the berries are pulled from their clusters. For this reason, when hand harvested, red and white currant fruit clusters are harvested intact when all of the fruit on a cluster is ripe. Black currants are firmer than red and white varieties and can be stripped from the clusters as individual berries ripen, or entire clusters may be picked by hand. As a practical matter, most currants are harvested in a single picking, or they may be harvested in two pickings one to two weeks apart. Gooseberries and jostaberries are picked when they are firm and come off as individual berries. These fruits typically ripen over a four-week period and are harvested in two to four handpickings. Harvest maturity for gooseberries varies greatly, according to the intended use (Ryall and Pentzer, 1982). Gooseberries that are not fully ripe (called green, regardless of the actual color) are very firm and tart and are used for processing. Fully ripe berries are soft, sweet, and well suited as fresh fruits and in desserts.

In countries where currants and gooseberries are grown on large scales, machine harvesting has largely replaced handpicking because of labor costs and shortages. Chapter 11 goes into detail regarding harvest maturity, harvesting methods, and storage practices.

Unlike strawberry and raspberry fruits that quickly go from ripe to overripe, ripe currants, gooseberries, and jostaberries can hang on the bushes for one to four weeks or more without becoming overripe. This characteristic provides growers with flexibility in harvesting and reduces the likelihood of losing part of a crop due to periods of bad weather during harvest.

## Fruit Composition

Currants, gooseberries, and jostaberries are rich in vitamins A, B, and C, potassium, phenolics, and flavanoids but low in calories and sodium (Tables 2.1 and 2.2). They provide moderate amounts of thiamin (vitamin $B_1$), niacin (vitamin $B_3$), and calcium. Black currant cultivars contain high concentrations of vitamin C, relative to other fruits, ranging from 0.05 to 0.25 percent (50 to 250 mg per 100 g fresh weight) (Brennan, 1996). Some wild black currant fruits have been found to contain up to 0.8 percent (800 mg/100 g fresh weight) of vitamin C. The amount of vitamin C in the fruit depends on cultivar, site, weather, and fertilization (Nilsson, 1969; Lenartowicz et al., 1976), although relative rankings between cultivars remain constant (Chrapkowska and Rogalinski, 1975). Among black currant cultivars, those descended from *R. dikuscha* contain the highest amounts of vitamin C (Gubenko et al., 1976). Cultivars derived from *R. nigrum* var. *sibiricum* have also been reported to contain high concentrations of vitamin C (Voscilko, 1969; Samorodova-Bianki et al., 1976; Voluznev and Zazulina, 1980). Hooper and Ayers (1950) determined that the vitamin C content of black currants is unusually stable due to the protective effects of fruit anthocyanins and flavones. Tamas (1968) reported that vitamin C stability was greater in cultivars derived from *R. nigrum* × *R. dikuscha* than from *R. nigrum* alone.

Unlike black currants, gooseberries and red and white currants contain relatively low concentrations of vitamin C, averaging around 40 mg/100 g (Zubeckis, 1962; Mapson, 1970). Chapter 12 suggests possible donors for increasing ascorbic acid concentrations in gooseberries and red currants.

In recent years, nutritionists have placed increasing emphasis on the importance of antioxidants in the diet as a means of reducing cancer, heart disease, and other health problems. Among the most common antioxidants are anthocyanins (common pigments in fruits) and various phenolic compounds. Black currants contain high concentrations of anthocyanins and total phenolics, and are high in antioxidant capacities (Moyer, Hummer, Wrolstad, and Finn, 2002; Moyer, Hummer, Finn, et al., 2002). According to Brennan (1996), developing cultivars with intense color, high ascorbic acid concentrations, and good flavor is a major breeding objective. Breeders today routinely use spectrophotometric and chromatographic techniques to

TABLE 2.1. Ripe fruit characteristics for currants, gooseberries, and jostaberries.

| Fruit | Berry weight (grams) | | | Average pH[b] | Average TSS (%)[c] |
|---|---|---|---|---|---|
| | average | minimum | maximum[a] | | |
| Black currants | 0.68 | 0.38 | 1.28 | 2.8 | 18.0 |
| Red currants | 0.68 | 0.42 | 1.06 | 2.8 | 11.7 |
| White currants | 0.60 | 0.52 | 0.66 | 2.9 | 12.5 |
| Gooseberries[d] | 3.75 | 1.17 | 7.37 | 3.1 | 13.0 |
| Jostaberry[e] | 2.31 | — | — | 3.0 | 14.5 |

*Source:* Adapted from Barney and Gerton (1992).
[a]Values represent the ranges in average berry weight of the cultivars evaluated.
[b]pH is a measure of the acidity of the fruit. A pH of 7.0 is neutral and values less than 7 are acidic.
[c]TSS = total soluble solids. This is a measure of sugar concentration in the berries.
[d]European and American gooseberries were evaluated as one group.
[e]Only one cultivar was available for evaluation.

TABLE 2.2. Nutritional value of currants, gooseberries, and other selected fruits (per 100 grams of fruit).

| Nutritional value | Black currants | Red currants | Gooseberry | Blueberry | Strawberry |
|---|---|---|---|---|---|
| Percent water | 84.2 | 85.7 | 88.9 | 83.2 | 89.9 |
| Calories | 54 | 50 | 39 | 62 | 37 |
| Protein (g) | 1.7 | 1.4 | 0.8 | 0.7 | 0.7 |
| Fat (g) | 0.1 | 0.2 | 0.2 | 0.5 | 0.5 |
| Carbohydrates (g) | 13.2 | 12.1 | 9.7 | 15.3 | 8.4 |
| Vitamins | | | | | |
| A | 230 | 120 | 290 | 100 | 60 |
| B (niacin) | 0.3 | 0.1 | — | 0.5 | 0.6 |
| $B_1$ (thiamin) | 0.05 | 0.04 | — | 0.03 | 0.03 |
| $B_2$ (riboflavin) | 0.05 | 0.05 | — | 0.06 | 0.07 |
| C (ascorbic acid) | 200 | 41 | 33 | 14 | 59 |
| Calcium (mg) | 60 | 32 | 18 | 15 | 21 |
| Phosphorus (mg) | 40 | 23 | 15 | 13 | 21 |
| Iron (mg) | 1.1 | 1.0 | 0.5 | 1.0 | 1.0 |
| Sodium (mg) | 3 | 2 | 1 | 1 | 1 |
| Potassium (mg) | 372 | 257 | 155 | 81 | 164 |

Source: Adapted from Watt and Merrill (1963).

evaluate prospective cultivars for their anthocyanin profiles and suitability for juice production. Table 2.3 lists anthocyanin, phenolic, and antioxidant characteristics for currants, gooseberries, blueberries, blackberries, and black raspberries.

In general, black currants compare favorably with blueberries, blackberries, and black raspberries in total anthocyanin and phenolic concentrations and in antioxidant capacities. Gooseberries and jostaberries contain low concentrations of these compounds and have low antioxidant capacities compared with black currants or the other fruits just mentioned. Given their lack of blue or purple pigments, red and white currants are also likely to rank lower than black currants in antioxidant activities, although data on these crops were unavailable.

It should be noted that antioxidant capacities and concentrations vary significantly among cultivars (Moyer, Hummer, Wrolstad, and Finn, 2002; Moyer, Hummer, Finn, et al., 2002). Of the 32 *R. nigrum* black currant cultivars tested, total anthocyanins ranged from 128 mg/100 g of fruit in 'Slitsa' to 411 mg/100 g in 'Consort' (Moyer, Hummer, Finn, et al., 2002). 'Consort' also had the greatest total phenolic concentration at 1342 mg/100 g and oxygen radical absorbing capacity (ORAC) at 93.1 µmol TE/g. 'Ben Conan' had the lowest total phenolic concentration at 498 mg/100 g and 'Kantata 50' the lowest ORAC at 36.9 µmol TE/g. Ferric reducing antioxidant power (FRAP) ranged from 61.5 µmol/g in 'Ojebyn' to 118.2 µmol/g in 'Kirovchanka'. The American black currant 'Crandall' *(R. aureum)* had anthocyanin concentrations, phenolic concentrations, and antioxidant characteristics similar to those found in European black currants *(R. nigrum)*. A Chilean black currant *(R. valdivianum)* was particularly rich in total phenolics and ranked only behind 'Consort' in total anthocyanins when compared with other tested black currants. *Ribes valdivianum* exhibited the greatest ORAC and FRAP capacities of all black currants. This species, however, has very small fruit that are not particularly palatable. Of all the species and cultivars tested, 'Munger' black raspberry contained the greatest total anthocyanin concentrations and *Ribes valdivianum* the highest total phenolic concentrations and FRAP values. 'Jewel' black raspberry exhibited the highest ORAC capacity.

Growers and processors are advised that cultivars should not be selected based solely on anthocyanin concentrations, ORAC, or similar criteria. Flavor, berry size, suitability for processing, yields, disease

TABLE 2.3. Average anthocyanin concentrations, phenolic concentrations, and antioxidant capacities in blackberries, blueberries, currants, gooseberries, jostaberries, and raspberries.

| Crop | ACY (mg/100g) | TPH (mg/100g) | ORAC (μmol TE/g) | FRAP (μmol/g) | ACY/TPH |
|---|---|---|---|---|---|
| Blackberry | 141 | 460 | 46.4 | 68.3 | 0.31 |
| Black currant | | | | | |
| R. aureum[a] | 273 | 958 | 68.0 | 107.8 | 0.28 |
| R. nigrum | 229 | 799 | 57.1 | 92.0 | 0.29 |
| R. valdivianum[a] | 358 | 1790 | 115.9 | 219.3 | 0.20 |
| Blueberry | | | | | |
| V. angustifolium[a] | 208 | 692 | 87.8 | 97.9 | 0.30 |
| V. ashei | 406 | 875 | 123.4 | 146.5 | 0.46 |
| V. corymbosum | 208 | 444 | 52.3 | 58.6 | 0.47 |
| Gooseberry[a] | 14 | 191 | 17.0 | 25.2 | 0.07 |
| Jostaberry | 74 | 309 | 28.1 | 39.0 | 0.24 |
| Raspberry, black | 566 | 955 | 117 | 186.3 | 0.60 |

Source: Adapted from Moyer and colleagues (Moyer, Hummer, Wrolstad, and Finn, 2002; Moyer, Hummer, Finn, et al., 2002).

Note: ACY = total anthocyanins; TPH = total phenolics; ORAC = oxygen radical absorbing capacity; FRAP = ferric reducing antioxidant power; ACY/TPH = ratio between total anthocyanins and total phenolics.
[a]Results based on tests with a single cultivar or genotype.

resistance, and other factors must also be determined. Caution is also required in comparing different crops. Moyer and colleagues (Moyer, Hummer, Wrolstad, and Finn, 2002), for example, noted that the anthocyanin composition of black currants was much less complex than those found in blueberries, blackberries, and raspberries. Each species and cultivar has a unique biochemical profile, and no single crop or cultivar is best for all uses.

Brennan (1996) summarizes some of the biochemistry research conducted on black currants:

> The anthocyanins contained in *R. nigrum* are mainly cyanidin and delphinidin 3-glucosides and 3-rutinosides (Harborne and Hall, 1964), with very small quantities of cyanidin-3-sophoroside, delphinidin 3-sophoroside, and pelargonidin 3-rutinoside, amounting to about 5% of the total (LeLous et al., 1975, cited by Brennan, 1996, p. 262).

Western European black currant cultivars primarily contain cyanidin derivatives, while those from Scandinavia are delphinidin derivatives (Taylor, 1989). By crossing western European and Scandinavian black currants, cultivars with high delphinidin:cyanidin ratios (such as 'Ben Lomond') have been produced (Brennan, 1996). 'Ben Lomond' served as a parent for the intensely colored 'Ben Alder', which is highly valued for juice processing. Although total pigment and delphinidin:cyanidin ratios are important, acylation is also an important consideration. Brennan (1996) noted that acylated anthocyanins retain their colors over a wide range of pH values and are less subject to deterioration than nonacylated pigments. Breeding efforts are underway to increase pigment acylation in black currant cultivars.

Pigmentation in gooseberries and red and white currants is somewhat simpler than in black currants. Although gooseberry fruit colors include white, green, yellow, red, and black, most European gooseberries contain only cyanidin 3-glucoside and cyanidin 3-rutinoside (Harborne and Hall, 1964). North American gooseberries may also contain delphinidin glycosides (Nilsson and Trajkovski, 1977). *Ribes robustum* and *R. niveum* were used in crosses with European gooseberries to produce black-fruited gooseberries. Color in red currants is produced by various forms of cyanidin 3-glycosides (Nilsson and Trajkovski, 1977; Oydvin, 1973).

Hukkanen and colleagues (1993) examined flavonol concentrations in ten black currant cultivars grown on conventional and organic farms. Although concentrations of the flavonols varied markedly between cultivars, no consistent differences were observed in the flavonol contents of the cultivars tested, regardless of whether they were grown conventionally or using organic methods. Myricetin was the most abundant flavonol, ranging from 8.9 to 24.5 mg per 100 grams of fresh berries. Quercetin levels also varied widely among cultivars, ranging from 5.2 to 12.2 mg per 100 grams of fruit. Kaempferol contents were low in black currants, lying between 0.9 and 2.3 mg per 100 grams of fresh fruit.

Flavor and aroma in black currants are highly complex, with at least 88 chemicals playing roles (Andersson and von Sydow, 1964, 1966; Nursten and Williams, 1967). The "foxy" or "catty" aroma associated with black currants appears to derive from sulfur-containing ketothiols (Lewis et al., 1980). Latrasse and Lantin (1974) correlated good flavor in several black currant cultivars with the presence of Δ-3 carene, β-phellandrene, and terpinolene in the bud essential oil. Flavor in red currants is simpler than in black currants, correlating with increased citric acid compared to tartaric and malic acids (Kronenberg, 1964).

### Dormancy

At about the same time the fruits ripen, *Ribes* begin acclimating for the coming winter. The first visible evidence of acclimation is the development of terminal buds on the tips of the new shoots. The shoots stop elongating and the soft, green stems thicken and begin to become hard and woody, a process called lignification. During lignification, the stems begin developing a layer of grayish bark.

*Ribes* leaves begin senescing during late summer and early fall. The color gradually changes from green to yellow, and the leaves dry out and fall. Leaves that have been damaged by powdery mildew, blister rust, or excessively high temperatures often turn grayish-brown and fall early with little or no yellowing. If defoliation occurs early in the year and temperatures and moisture are favorable, the buds can break and begin growing. The succulent growth that develops does not acclimate properly and can be killed by fall frosts. Red currants enter dormancy earlier than black currants (Mage, 1976).

Once the canes enter full dormancy, they must go through a period of chilling, as described earlier, before they can begin growing again.

During the dormant period, gooseberries and currants are among the most cold-hardy fruit crops. Fully acclimated canes in deep rest can tolerate winter temperatures of –22 to –31°F (–30 to –35°C) or colder without injury (Harmat et al., 1990), and these crops are widely grown in northern latitudes where temperatures of –40°F (–40°C) or lower are common. At the time they break in the spring, buds are injured by temperatures between 19 and 23°F (–7 and –5°C). Flowers and developing fruit are killed or injured at temperatures between about 28 and 31°F (–2 and 0°C).

## Roots

Many domesticated *Ribes* develop vigorous root systems. Plants propagated from seed produce taproots that can penetrate the soil to depths of 39 inches (1 m) or more. Most cultivated gooseberries and currants, however, are propagated as rooted cuttings. *Ribes* do not breed true from seed, and asexual propagation methods, such as rooted cuttings and layering, are used to produce true-to-name cultivars. Plants propagated from cuttings or layers form shallow adventitious roots that generally grow in the top 16 inches (40 cm) of the soil and eventually spread out to form a circle about 32 inches (80 cm) in diameter (Harmat et al., 1990). Tall, vigorous bushes can develop larger root systems. Roots near the soil surface are especially abundant in root hairs, one reason why mulching, herbicides, and/or shallow cultivation are preferred over deep cultivation for weed control.

Although the tendency to quickly form vigorous adventitious roots makes propagating most currants and gooseberries easy, it makes long-term container culture difficult. Currants, gooseberries, and jostaberries planted as rooted cuttings into No. 1 nursery pots (approximately one gallon) often become rootbound within a year.

# Chapter 3

# Propagation

## *SEEDS AND STOCK*

*Ribes* seeds are long-lived. The seeds from most species have a dormancy requirement that can be broken by cold, moist conditions in a process called stratification. The following method has proven effective in germinating most *Ribes* seeds. Bear in mind that seeds represent sexual reproduction. Seedlings are genetically different from their parents and may or may not prove suitable for fruit production. Fruit breeders are most likely to use seed propagation in developing new cultivars, while currants and gooseberries intended for commercial production are seldom propagated from seed.

Seed extraction and storage techniques are described in Chapter 12. Seeds that have been properly extracted, dried, and stored in a desiccator at −4°F (−20°C) normally remain viable for several years (Brennan, 1996).

1. Surface sterilize the seeds with a solution containing two to three fluid ounces per quart (75 to 100 ml/L) of sodium hypochlorite-based household bleach for five minutes. After sterilization, rinse the seeds in two or three cool, tap-water baths for fifteen minutes to remove the bleach.
2. Stratify the seeds at 35°F (2°C) for eight to twelve weeks on moist sand or paper towel.
3. Germinate the seeds using temperatures of 60°F (16°C) for 14 hours alternating with 72°F (22°C) for 10 hours. Expose the seeds to light during the warm period.

*Ribes* are usually easy to propagate by layering canes or rooting cuttings. Propagating vegetatively (asexually) in these ways produces plants genetically identical to the original plants from which they were derived. For consistent quality, growers use vegetatively propagated clones, called cultivars, whose characteristics are known.

Before vegetatively propagating any cultivars, ensure that you may legally do so. Plant breeders' rights legally protect many modern cultivars. These rights serve as patents and restrict propagation to nurseries licensed to propagate those specific cultivars.

Although vegetative propagation is usually easy, most commercial growers purchase stock from nurseries. Quality is a critical factor in propagating stock, and one must ensure that the plants are kept free from pests and diseases. Agricultural agencies inspect commercial nurseries and certify that they are taking steps to produce clean stock. Unless safeguards are taken, insects, mites, viruses, and diseases can be introduced on contaminated plants. Growers who propagate their own stock should choose material from vigorous, healthy, pest-free mother plants and take stringent precautions to control pests and diseases in propagation beds.

Currants, gooseberries, and jostaberries are typically propagated by layers or cuttings. It takes one to two seasons to grow a cutting or layer to planting size, depending upon the cultivar. Vigorous, easy-to-root cultivars propagated in the spring should be ready to plant by the first fall. European gooseberries are sometimes more difficult to propagate than other *Ribes* and may require two years to reach planting size.


## *LAYERING*

Layering is usually done in the early spring to produce small amounts of planting stock. To produce a one-rooted layer, bend a low-lying, one-year-old cane to the bottom of a shallow hole next to the mother plant and pin the cane into place with a U-shaped piece of wire, notched stick, or weight. Bend the cane sharply where it is pinned to the ground and make a small notch in or girdle the bark at the bend to stimulate root formation. You will not need to treat the layer with a rooting compound. Fill the hole with soil and make a soil mound four to six inches (10 to 15 cm) high, leaving the tip of the cane exposed. After the plant becomes dormant in the fall, cut the rooted layer from the mother plant, taking care not to damage the new plant's roots. Keep as much soil around the roots as possible. The rooted layer can be transplanted into a transplant bed or production field immediately or stored at temperatures near freezing until planting. If the layers are to be stored in a refrigerator, soak the roots with water and enclose the layers in plastic bags filled with moist sawdust.

A variation of simple layering, called stooling or mound layering, can be used to produce larger numbers of plants. Start by cutting off all of the canes on an established bush one to two inches (3 to 5 cm) above the ground after the plants become dormant in the fall or before growth begins in the spring. After the new shoots have grown about six inches (15 cm) tall, mound sawdust or compost over them, leaving several inches of the shoots exposed. Work the mulch in around the canes with a shovel or other implement to eliminate air pockets. Add more sawdust or compost when the shoots are about twelve inches (30 cm) tall. Irrigate frequently to prevent the mounds from drying out. Although soil can be used instead of sawdust or compost, mounding with the organic mulches allows easier removal of the rooted layers. When layering in a stool bed, do not bend, pin, or wound the shoots before mounding. Roots form readily on the young shoots, and the layers can usually be cut from the stool bed that fall and planted for an additional year in a nursery transplant bed. European gooseberries may require two growing seasons in the stool bed (Card, 1907; Van Meter, 1928). New shoots will form from the cane stubs and crowns in stool beds, which can remain productive for years. Simple layering and stooling are illustrated in Figure 3.1.

## CUTTINGS

Both hardwood and semi-hardwood cuttings can be rooted with relative ease. Collect dormant hardwood cuttings from late autumn after the leaves fall through late winter before any new growth starts. Four- to eight-inch (10 to 20 cm) cuttings from wood that grew during the previous summer work well. Older references recommend that hardwood cuttings be a foot long (Card, 1907), but that practice has fallen out of favor. Make a flat cut on the base of the cuttings (closest to the mother plant) and a slanting cut on the tip end to help identify the bases and tips of cuttings. Make the top cut about one-quarter inch (6 mm) above a bud and bundle the cuttings together in groups of about 25. Store the bundles buried upside down either in the ground with at least two inches (1 cm) of soil covering the butts of the

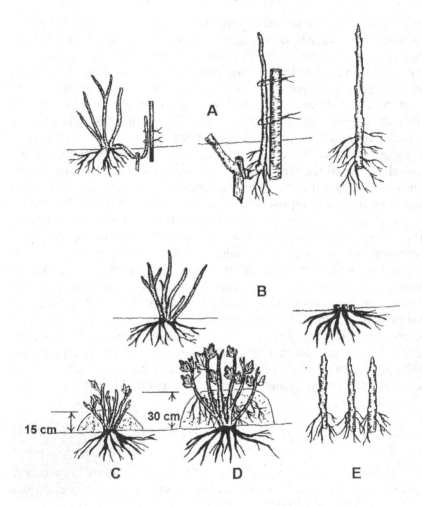

FIGURE 3.1. Two types of propagation. (A) Propagation using simple layering. While the plant is dormant in the spring, peg a single one-year-old cane to the bottom of a six-inch (15 cm) deep hole, tying the tip of the shoot to a stake above ground. Wound the cane by removing a strip of bark at the end. (B-E) Propagation using mound layering or stooling. (B) Prune off all canes from dormant bushes, leaving one- to two-inch (3 to 5 cm) stubs above ground. (C) When new shoots are six inches (15 cm) tall, mound three to four inches (8 to 10 cm) of sawdust around the shoots. Do not wound the shoots. (D) When the shoots are 12 inches (30 cm) tall, add four to six inches (10 to 15 cm) more sawdust. (E) Dig the rooted layers in the fall.

cuttings or inside a cool cellar in moist sand or sawdust at about 32 to 34°F (0 to 1°C) until planting in the spring (Card, 1907; Van Meter, 1928). Overwintering cuttings this way encourages callus development on the cut surfaces.

An alternative to burying hardwood cuttings is to wrap them in moist (not wet) paper towels, place them in a plastic bag, and store them in a refrigerator (not freezer). Dipping or spraying the cuttings with a broad-spectrum fungicide before storage helps prevent mold. Rather than overwintering cuttings, some growers in the northeastern United States prefer to harvest hardwood cuttings from mid-September to mid-October, soak the bases of the cuttings in water for one to two weeks, and stick the cuttings directly into nursery beds.

Traditionally, dormant hardwood cuttings were stuck directly into nursery beds in the early spring (Auchter and Knapp, 1937; Van Meter, 1928). The cuttings were spaced four to six inches (10 to 15 cm) apart and buried base down with two buds left above the soil surface. Although this method works well for American gooseberries and most currants, European gooseberries are sometimes harder to root and respond better to either stool layering (described earlier) or semi-hardwood cuttings. Rather than use a nursery bed, some propagators prefer to root cuttings in containers, which gives more flexibility in harvesting, shipping, and transplanting the cuttings after they root.

Semi-hardwood cuttings (sometimes called softwood cuttings) are collected from the current season's growth during June or July. Irrigate mother plants several days before collecting the cuttings to avoid drought stress. Collecting cuttings early in the morning also helps ensure they are properly hydrated.

Select stem sections on which the leaves have reached full size and the stems have begun to harden. Discard the succulent shoot tips, which dry out easily. Make the cuttings four to eight inches (10 to 20 cm) long, ensuring that each cutting has at least four buds. The cuttings should be about the thickness of a pencil. Leave two or three leaves at the tops of the cuttings and strip off all other leaves. Cut off the outer one-half of the retained leaves to reduce moisture loss and enclose the cuttings in plastic bags. Unless the cuttings will be potted immediately, put them into a refrigerator, but do not allow them to freeze.

For best results, dip the bottoms of hardwood and semi-hardwood cuttings into a dry or liquid rooting compound containing from 1000 to 3000 parts per million (ppm) indole butyric acid (IBA). Rooting hormones are especially useful for propagating hard-to-root European gooseberries. Rooting compounds are available through commercial horticultural suppliers and garden centers. Always follow label directions.

Whereas hardwood cuttings are often stuck directly into nursery rows, semi-hardwood cuttings are more delicate and susceptible to drying out and are generally easier to root in containers or raised nursery beds. Some propagators prefer to root cuttings in small plastic pots approximately four inches (10 cm) square to conserve potting soil. Once the pots are filled with roots, the cuttings are transplanted to nursery beds for growing out. If weather conditions are favorable and the cuttings are large enough and well rooted, they can be transplanted directly to production fields.

Before using them to root cuttings, wash used containers with soap and water and sterilize them to kill disease-causing organisms. Immersing the containers for 30 minutes in a solution of one part household bleach and nine parts water kills most disease organisms. Rinse the containers thoroughly in clean, fresh water after sterilizing to remove traces of the bleach.

Soil in nursery beds should be well drained but not droughty. As long as these criteria are met, native soils, amended soils, and soilless potting soils all work well for rooting *Ribes*. Field and garden soils seldom work well for container culture. For rooting in containers, fill the pots with a mixture of 50 percent potting soil, peat moss, or well-rotted compost and 50 percent perlite or sand. Do not add fertilizers.

Stick the cuttings into the nursery beds or pots, ensuring the cuttings are right side up. If you treat the cuttings with a dry rooting compound, stick the cuttings into preformed holes to avoid wiping the rooting compound off on the soil. Leave one or two buds exposed above the media.

Screening the cuttings from direct sunlight can help prevent them from drying out. Keep the soil continually moist but not waterlogged. Beginning immediately for semi-hardwood cuttings, and when leaves begin to appear on hardwood cuttings, mist with water several times daily.

Growers often find it easier to root softwood and semi-hardwood cuttings in either a misting bed or under a tent of clear polyethylene. If mold becomes a problem, treat the cuttings with a fungicide. Always follow label directions when using fungicides and other pesticides. Bottom heat set at 70°F (21°C) improves rooting. Heating cables and pads are available through greenhouse supply companies and can be used in nursery beds or under containers.

Adventitious roots normally form within about four weeks. Resist the temptation to pull cuttings out of the beds or pots to check for rooting. Doing so will damage delicate new roots and delay rooting and growth. Roots growing out of the bottoms of the containers indicate successful rooting. Flushes of new shoots are also a good indication that cuttings have rooted.

Once the cuttings are rooted, begin adding slow-release or liquid fertilizers, according to label directions. Harden off the cuttings over a two-week period by gradually increasing their exposure to sunlight before placing them in full sun. *Ribes* need to undergo a chilling period during the winter, so do not grow them indoors to extend their growing season.

Currants and gooseberries are sometimes propagated by tissue culture. Unlike raspberries and strawberries, which are often available from commercial nurseries in tissue culture form, *Ribes* are generally tissue cultured only for research purposes or to eliminate virus contamination.

# Chapter 4

# Selecting a Planting Site

Selecting a site is a critical decision a grower faces in establishing a berry farm. Purchasing land and preparing it for farming are long-term decisions that are difficult and expensive to change. Although many factors must be considered when selecting a farm site, they can be divided into two main categories: nonsite considerations and site considerations. Nonsite factors include the intended crop, how and where it will be marketed, whether the fruit will be processed before marketing, and the amount of capital available. Site considerations include climate, soil, topography, winds, endemic pests and diseases, and access. Appendix A provides a checklist for evaluating these factors.

## *NONSITE CONSIDERATIONS*

How will the crop be used? If currants, gooseberries, or jostaberries are to be marketed fresh, several factors must be taken into consideration. The site must allow the grower to get his or her fruit to market in a timely fashion.

Harvesting is the largest single recurring expense in *Ribes* production, even when labor is readily available. To reduce production costs, some growers elect to establish U-pick or pick-your-own farms. Most successful U-pick enterprises are located within about 15 miles (24 km) of a city or large town and are both visible and easily accessible to customers.

If the fruit is to be processed in some way, make sure you can legally do the processing and that you have adequate processing facilities available in your area. If your primary interest is in making jams, jellies, or other value-added products, you may find it more economical to devote your time and resources to processing and let someone else grow the fruit.

## SITE CONSIDERATIONS

### Climate

There is an old axiom in farming: Climate sets the limits within which all other factors operate; if the climate is wrong for a particular crop, no amount of effort will allow economical production of that crop. One must consider the long-term climatic history for a planting site, as opposed to short-term weather patterns over one or a few years. For currants, gooseberries, and jostaberries, the major concerns are summer temperatures, the amount of chilling plants receive during the winter, spring frosts, and length of the growing season. Low winter temperatures are seldom a problem for *Ribes* growers in the northern United States or southern Canada.

### Summer Temperatures

Although currants and gooseberries are renowned for their cold-hardiness, they do not tolerate high temperatures well, especially when combined with intense sunlight. Research indicates that leaf tissues can be injured at temperatures as low as 86°F (30°C). After a month of daily temperatures in the 90 to 105°F (32 to 41°C) range, combined with clear weather, one of the authors noted significant leaf damage and sunburned fruit on gooseberries and currants in irrigated research plots in Idaho. The jostaberry cultivar Josta, however, exhibited little or no damage.

*Ribes* perform reasonably well in partial shade, and interplanting them between fruit trees in an orchard was once a common practice in North America. Exposure to full sun, however, is generally required for maximum yields. In warmer climates, growers can plant currants and gooseberries in partial shade to reduce heat stress. High elevation sites and north-facing slopes are generally preferable in warmer climates (Sears, 1925).

### Fall and Winter Temperatures

As discussed in Chapter 2, *Ribes* require a dormant period during the winter in order to flower and grow normally. In late summer and early fall, short days and cool temperatures trigger the onset of winter acclimation and dormancy. Once plants are fully dormant, they require approximately 800 to 1600 hours exposure to temperatures between 32 and 45°F in order to complete their chilling requirements and begin growth (Westwood, 1978). Short, warm winters can prevent *Ribes* crops from meeting their dormancy requirements. European gooseberries are, reportedly, somewhat more tolerant of low chilling conditions than other cultivated gooseberries, currants, and jostaberries.

Although they do not tolerate high temperatures, many gooseberry and currant cultivars are very cold hardy (generally considered more cold hardy than apple trees). Cultivars differ in their hardiness, however, and some are hardy only to about –13°F (–25°C) (Holubowicz and Bojar, 1982). Unfortunately, coldhardiness data are lacking for many cultivars. Little has been published on jostaberry coldhardiness.

In cold, windy areas, windbreaks improve *Ribes* survival and performance by reducing desiccation damage to the canes. Winter cold-hardiness is not an absolute value for any cultivar but varies according to weather conditions, moisture availability, soil fertility, and crop load during the previous summer and fall, as well as the weather conditions and time of year. Plants stressed by heat, drought, disease, pests, poor nutrition, or overcropping may not acclimate fully and are likely to suffer winter injury.

The problems with heat damage and the difficulty in meeting dormancy requirements make currant, gooseberry, and jostaberry cultivation difficult in southern climates. Figure 4.1 shows areas of North America where cultivated *Ribes* are adapted. Before planting, however, growers must ensure they can legally produce these crops in their areas. Table 1.1 summarizes state restrictions on importing and growing currants and gooseberries.

## Spring Frosts and Length of the Growing Season

The time between the last frost of spring and the first frost of fall is a major consideration in selecting a site, since many actively growing plants cannot tolerate temperatures below about 28°F (–2°C). *Ribes* are best grown in areas with frost-free periods of 120 to 150 days. Marginal sites may require overhead sprinklers, wind machines, or other devices to prevent spring frost damage. The cost of such frost protection can be high.

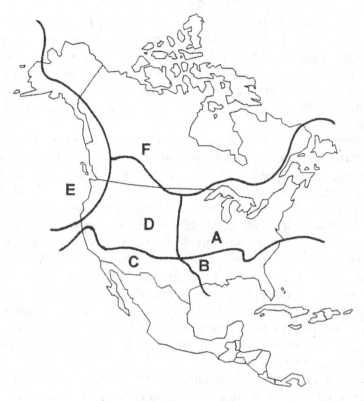

FIGURE 4.1. Map of North America showing suitable climates for commercial *Ribes* production. Domesticated currants, gooseberries, and jostaberries are well adapted to regions A and E. In region B, the summers tend to be too long and hot, and lack of winter chilling can become a problem. Area C is too hot and dry for reliable commercial *Ribes* production. Relatively low summer rainfall limits dryland production in region D, but *Ribes* can be grown with irrigation. Region F provides some growing opportunities, but winter injury, spring frosts, and short growing seasons may create difficulties. Note that these regions are not sharply defined but gradually blend into one another depending upon elevation and other factors. (*Source:* Adapted from Darrow, 1919.)

*Ribes* are among the earliest-blooming domestic fruits and are susceptible to spring frosts. Blossoms are killed or injured by temperatures between about 28 and 31°F (−2 and 0°C) (Harmat et al., 1990). To reduce frost damage, select a planting site with good cold air drainage, as discussed later in this chapter.

Growers can take steps to protect their crops on sites susceptible to spring frosts. On large fruit farms, growers in frosty areas often install wind machines to break up temperature inversions and keep freezing air from settling into the fields. Helicopters are sometimes flown over fruit fields during spring frosts for the same purpose. Both methods are usually too expensive for small-scale berry farmers. At one time, fruit growers protected fruit blossoms from frost by burning tires and smudge pots in their fields. This strategy is now illegal in some regions because of the large amounts of smoke produced.

One method of frost protection that can be adapted to many sites uses irrigation water to prevent blossoms from freezing. Growers using this method begin sprinkling irrigation water over their crops when the air temperature drops to 32°F (0°C). At least one-quarter inch (6 mm) of water per hour is applied to the bushes. Although ice forms on the plants, the blossoms are generally not damaged as long as liquid water is present. Sprinkling continues until the air temperature rises above freezing and the ice begins to melt. The blossoms are not damaged because water releases heat as it freezes, keeping the ice and blossoms around 32°F. If sprinkling stops while the air temperature is below freezing, the temperature of the blossoms will drop and they may be damaged. On frosty sites, prospective *Ribes* growers may wish to include the cost of a sprinkler irrigation system in enterprise budget calculations. They must also ensure that their sites provide adequate irrigation water capacity to apply water throughout frosts.

## Water

Currants and gooseberries are moderate in their water consumption, typically using about one inch (3 cm) of water per week during the growing season in the northern United States and southern Canada. Prolonged wet weather during the early spring, especially when combined with cold temperatures, can create serious disease problems. Wet weather also discourages bee activity and can interfere with pollination. Because of their stout, straight, lightly branched canes, *Ribes* are seldom injured by heavy snow. In fact, prolonged snow cover during the winter helps to insulate the plants from severe winter temperatures.

Most growers in the United States and Canada find irrigation beneficial. When evaluating a site, determine what sources of water are available and what water rights apply. Wells, springs, streams, lakes, and irrigation ponds are all used for irrigation. Determine if the water is seasonal or available year-round, what the capacity of the source is, who has legal access to the water, and how much accessing the water will cost.

When evaluating a potential site, consider what type of irrigation system will be used. Furrow irrigation (also called rill or flood), overhead sprinklers, and trickle or drip irrigation systems are all used in fruit-growing operations. Irrigation systems are discussed in more detail in Chapter 5.

### Wetlands and Adjacent Surface Waters

Another factor to consider in terms of water is whether lands are legally classified as wetlands or highly erodible. If so, use of the site may be regulated and unavailable for cultivation. Before purchasing any site in the United States, check with the U.S. Department of Agriculture Farm Service Agency to determine if the site is regulated. In Canada, check with the local Agriculture Canada office.

Adjacent surface waters can be both assets and liabilities. If one can gain access to the water for irrigation, the water is an asset. Legal requirements to protect the surface water, however, may have to be met.

### Soil

Contrary to popular opinion, soil fertility is a relatively minor consideration in selecting a site. Whereas improving fertility is generally easy, changing other soil characteristics can be difficult. Correcting drainage problems on a wet site, improving water-holding capacity on a droughty site, and making major changes in soil pH and organic matter are difficult and often expensive.

*Ribes* tolerate a wide range of soils but perform best on deep, cool, well-drained loams with good moisture-holding capacity and large amounts of organic matter (Shoemaker, 1948). Some authorities have recommended planting currants and gooseberries on heavy soils containing large amounts of clay (Card, 1907; Sears, 1925; Van Meter, 1928). Although *Ribes* are less susceptible to root diseases than raspberries and some other berry crops, they should not be planted on poorly drained sites (Keeble and Rawes, 1948; Van Meter, 1928).

Many horticulturists during the 1800s and early 1900s noted that *Ribes* crops survived on heavy soils but that yields were reduced.

A water table between three and six feet (1 to 2 m) deep is best. Water table depths less than three feet can create root problems (Harmat et al., 1990). Poor drainage is usually associated with large amounts of clay in the soil, high water tables, and hardpans. Such problems are often indicated by very dark soils that are high in organic matter or have gray-green streaks in their profile. Sandy or otherwise droughty soils can create problems with heat damage and premature defoliation (Shoemaker, 1948). Soil surveys are available for many agricultural areas. The surveys, often available from local agricultural agents, give detailed information on soil types within an area, along with soil characteristics and recommended management practices.

Although published soil surveys provide valuable information, each site must be evaluated on its own merits. One should never buy a site for farming unless he or she has taken representative soil samples and has a clear picture of what the soil profile looks like. Be especially careful in determining the depth of the topsoil and where the water table is throughout the year. Determine whether there are hardpans or clay layers and how deep they are. Inspect the soil profile, looking for indications of drainage or drought problems. Even a small site may have several different soil types, not all of which might be suitable for currants, gooseberries, or jostaberries.

A soil analytical laboratory should test soil from each potential site. The analyses provide information on the types of soil, pH, fertility, and potential salinity problems. Once crops are established, fertilizer recommendations are typically based on foliar analyses of leaves and petioles. For those unfamiliar with soil testing, a representative from an analytical laboratory, county cooperative extension educator, or Agriculture Canada representative should be able to provide guidelines on how to collect samples. At a minimum, the laboratory should test for pH, exchange acidity, salinity, nitrogen (N), phosphorus (P), potassium (K), sulfur (S), calcium (Ca), boron (B), and sodium (Na). Most soil labs provide fertilizer and liming recommendations based on the analyses, if requested.

*Ribes* should have at least 1,000 parts per million of calcium in the soil and can tolerate up to 150,000 ppm (15 percent) calcium. Total salt and sodium concentrations in the soil should not exceed 15,000 ppm and 500 ppm, respectively (Harmat et al., 1990). A pH value of 7.0 is neutral. Soils with pH values below 7.0 are acidic, while those with values above 7.0 are alkaline or basic. For *Ribes* production, the soil pH levels should lie between 5.5 and 7.0. Soil pH is important because it affects the availability of mineral nutrients. If, for example, the soil pH is much above 7.0 and the soil is high in calcium, iron chlorosis may restrict plant growth. Chapter 6 discusses strategies for adjusting soil pH, if needed.

### Topography and Site Surroundings

The slope, orientation, and surroundings of a site have a profound impact upon its suitability for fruit production. Factors to consider include air drainage, light exposure, compass orientation, proximity to large bodies of water, proximity to other farms, air pollution, slope, and winds.

### Air Drainage

Adequate air drainage is important in minimizing frost damage, freezing injury, and disease problems. Low-lying frost pockets are poor locations for *Ribes* crops. Far better are benches and gentle slopes lying above cold-air inversion layers. Select fields that allow heavy, cold air to drain downhill, unimpeded by windbreaks, hedgerows, elevated roadways, or other obstructions.

Sites that trap moist air and restrict air movement within and out of fields delay drying of foliage and fruit after rains or irrigation. As a result, bacterial and fungal disease problems increase greatly.

Sites must be level enough to accommodate fruit growing. Although a slope can improve air and water drainage, steep slopes increase erosion and interfere with cultural operations. Steep terrain also creates serious safety hazards when tractors or other machines are used.

### Light Exposure

Sunlight provides the energy that green plants require for growth and survival. Anything that limits light exposure, such as shading by mountains, trees, or buildings, reduces the energy available to plants. Although currants and gooseberries tolerate partial shade, maximum yields usually require a full-sun setting. In hot climates, partial shading, especially in the afternoon, can help reduce leaf and fruit damage in *Ribes* crops.

## Compass Orientation

Sites with southern or western exposures warm more quickly in the spring and are warmer throughout the year than northern or eastern exposures. Warm, early spring temperatures not only encourage earlier growth and flowering, which are desirable for some crops, but also increase the potential for frost damage. North-facing slopes are generally recommended for *Ribes* production because they are cooler than south- or west-facing slopes, thereby retarding bloom in the spring and reducing the risk of frost damage (Shoemaker, 1948; Sears, 1925; Van Meter, 1928).

## Proximity to Large Bodies of Water

Open water is cooler than surrounding land in the spring and warmer during the fall and winter. Because of the moderating influence that large bodies of water have on air temperatures, coastal areas, even in the north, are characterized by mild climates and long growing seasons. Cool air blowing in from the ocean or large lakes delays spring blooming and fall frosts, thereby reducing the likelihood of frost damage. It is the reduction in spring frost damage that benefits *Ribes* growers near large bodies of water. Once water freezes, however, any moderating influence it has on temperature is lost.

Because prevailing winds in the northern hemisphere move from west to east, sites along eastern shores experience a greater moderating effect upon temperatures than those on western shores. The influence a body of water has on surrounding temperatures is directly dependent upon its size. The ocean has a large impact on the climate for hundreds of miles inland. Lakes generally have little influence upon temperatures more than a few hundred yards to a few miles away. Unless they serve as air drainages from warm areas, rivers seldom affect air temperatures more than a few hundred yards from the water.

## Proximity to Other Farming Operations

When evaluating a site, consider nearby farming operations. Pesticide vapors, spray drift, and surface runoff water contaminated with herbicides and other pesticides can kill or injure susceptible plants and make crops unmarketable. Contamination from other sites can make it difficult or impossible to maintain organic farming certification standards. Pests, weeds, and diseases can also spread from adjacent fields, woods, and windbreaks.

*Proximity to Air Pollution Sources*

Crops can be adversely affected by air pollution. Pollutants often associated with crop damage include ozone, sulfur dioxide, nitrogen oxides, peroxyacyl nitrate, and fluorides. Smoke and smog impact plants indirectly by reducing light intensity and subsequent photosynthesis. Little information is available, however, concerning the effects of air pollution on *Ribes* production. Considering that these crops are widely grown throughout heavily industrialized countries in Eastern Europe, it seems reasonable to conclude that cultivated *Ribes* are not particularly susceptible to air pollution.

*Winds*

In evaluating a site, consider the direction, frequency, duration, and velocity of prevailing winds. In general, currants, gooseberries, and jostaberries are not particularly susceptible to wind damage (Harmat et al., 1990) but in windy areas should have some protection from cold winds (Keeble and Rawes, 1948). In cold areas where snow cover is often limited, such as the Great Plains, cold, dry winds during the winter can damage canes through desiccation. Windbreaks may be beneficial on such sites.

### Pests and Diseases

Proper site and crop selection can greatly reduce pest and disease problems, and one should find out about the pests, weeds, and diseases endemic to his or her area. Pay particular attention to the *Ribes* pests and diseases described in Chapter 10.

Avoid planting black currant cultivars not resistant to white pine blister rust on sites near white, sugar, bristlecone, whitebark, or any other five-needled pines. Sites surrounded by tall trees are poor locations for berry crops because birds roost in the nearby trees and feed on the berries. Birds will feed on any *Ribes* fruits, with red-fruited currants and gooseberries being, perhaps, the most attractive.

Throughout North America, predation from deer, moose, or elk causes substantial crop losses. Berry crops, in general, are susceptible to browse damage, even in populated areas. *Ribes* are moderately attractive to deer and moose. In Idaho test plots, unprotected red, white, and black currants; gooseberries; and jostaberries suffered little or no damage during most seasons, compared with adjacent grapes, raspberries, blackberries, saskatoons, apples, pears, plums, and cherries that were moderately to severely browsed by white tail deer and/or moose. When damage did occur on the *Ribes* crops, jostaberries were the most frequent target. Browse damage can increase dramatically during years when deep snow or other factors create food shortages for deer and moose. New shoots and leaves are the preferred food, and small, newly transplanted bushes are most at risk from browse damage. Dormant shoots may also be browsed during the winter. Herbivore pest control is discussed in Chapter 5.

A frequently neglected factor in site selection is toxic plants. Poisonous weeds are a special concern in U-pick operations, but they present dangers to all farm workers and visitors. Several nightshade *(Solanum)* species, for example, produce toxic red to black berries that could be mistaken for currants. Find out about dangerous plants in the area and look for them when evaluating sites. Most weed species can be eliminated before planting on an otherwise acceptable site. Regular scouting is an important step in keeping fields free of potentially dangerous plants.

### History of the Site

Previous pesticide use can have consequences, especially for farmers attempting to meet organic certification standards. During the early twentieth century, mercury- and arsenic-based pesticides were commonly used, especially in orchards. Although now banned, those pesticides are persistent in the soil and can contaminate present crops or make organic certification impossible. Some herbicides are persistent in the soil and can create production problems for years.

## Roads and Access

Road access is critical to all farming operations. When shipping berries, a paved road will give a smoother, cleaner ride than a gravel road. Adjacent dirt or gravel roads can create dust contamination on fruit. In evaluating a site, consider the following questions:

- When will access be needed to the site?
- Will local roads and highways be accessible when needed?
- Are the roads suitable for farm vehicles and products?
- Who is responsible for maintaining the roads?
- Who pays for road maintenance, and what is the cost?
- Will farming operation damage the roads and adjacent areas?
- Who provides weed control along adjacent roads?
- Will weed control measures along the road adversely affect *Ribes* production and marketing?
- Will residents or businesses be disturbed by or complain about traffic to and from the site?
- Can products be safely and economically transported to market?
- Is safe, adequate, and convenient parking available for employees and customers?

## Marketing

When fruit is to be marketed directly through a roadside stand or on-farm sales, the site should be in a highly visible location that is easily accessible from a main thoroughfare. Provide ample and convenient parking. Before opening a stand, determine if there is sufficient traffic in the right season and at the right times of day to support the stand. Is the traffic made up of people who are likely to stop and buy the produce and products? Ensure potential customers can see the stand far enough ahead to stop safely and comfortably. Evaluate the site critically. Is it conducive to shopping, or do junkyards, dilapidated buildings, or malodorous industries surround it?

## Utilities

Will the farm require electricity or gas to power pumps, refrigerators, heaters, lights, or other equipment? Will a telephone be needed? How much will installation and use of utilities cost? Will the enterprise generate waste products that will have to be removed by a commercial firm? These are all questions that should be answered when evaluating sites.

## Support Services

Successful farming operations today are complex and generally require interaction with many other businesses and agencies. If other fruit growers are in the area, it may be possible to form cooperatives to facilitate purchases and marketing. Established farmers are also good sources of technical information dealing with production and marketing. Ensure that supplies, equipment, and equipment service are available. If the fruit is intended for processing, determine if businesses in the area can process it. Growers planning to process their own fruit should determine if commercial kitchen facilities are available.

## Labor

Fruit farmers require skilled labor to produce and harvest their crops. Because currant, gooseberry, and jostaberry production in North America is very limited, workers experienced with the crops are seldom available. Determine how much labor will be needed and when. Will workers need special skills? If so, can unskilled laborers be trained adequately to carry out necessary operations? Who will provide the training and how much will it cost?

Migrant farm workers are an important part of commercial agriculture in much of North America but are often in short supply in nontraditional growing regions. If migrant workers are needed, how will they be contacted? In some cases, migrant labor brokers may be available. If migrant workers are available, however, one must still consider how they will be housed and transported. Strict government regulations pertain to hiring migrant workers, the first being that the workers are in the country legally. Housing requirements are also specific.

## Legal Considerations

Local, state, provincial, and federal regulations set guidelines for and limits upon many farming enterprises. Find out about applicable regulations before beginning any operation. Some counties in the United States, for example, are zoned primarily as residential and limit on-farm and roadside sales and related activities. Ensure that zoning regulations allow you to carry out planned operations. Allow for future growth and diversification. Beware of establishing yourself too near city boundaries. In recent years, cities have been able to annex farmland against the owners' objections. Trying to run a farm within city limits can be a difficult or impossible task.

Find out what taxes will be assessed at prospective locations and whether tax incentives are available for new businesses. Some communities prohibit the use of certain equipment (trailer houses, for example) or require that certain improvements be made on property. Find out in advance what will be allowed, what will be required, and what the costs will be.

Before purchasing property, consult with a local attorney. Determine if there are any liens or restrictive covenants on prospective sites. Ensure that you can do all that you want and need to do in connection with your intended enterprise.

# Chapter 5

# Designing a Currant, Gooseberry, or Jostaberry Farm

Along with selecting a site and crops, site design is a critical step. Irrigation systems, fences, roads, and farm buildings are expensive to establish and often both difficult and expensive to change. Laying out a farm properly saves time, effort, and money in the long run. Create a design that will optimize (not necessarily maximize) yields and facilitate cultural practices, such as mowing cover crops, spraying, pruning, and harvesting.

## FIELD LAYOUT

Recommendations for the distance between currant and gooseberry rows vary from five to twelve feet (1.2 to 3.5 m) (Card, 1907; Harmat et al., 1990). During the nineteenth and early twentieth centuries when horses were used to cultivate between rows, spacings of about six feet (2 m) were common. One of the authors has maintained a currant, gooseberry, and jostaberry plantation with rows spaced eight feet (2.5 m) apart and found it difficult, even when using small tractors, to avoid damaging the bushes. Although some growers prefer closer spacing, rows should generally be spaced at least eight and preferably ten feet (2.5 to 3 m) apart for currants and gooseberries, and ten to twelve feet (3 to 3.5 m) apart for jostaberries or when large tractors are used. Although closer spacings increase yields when the bushes are young, overcrowding and decreased yields will occur as the plantation matures over a five- to seven-year period. Install crossroads at right angles to the crop rows and no more than 300 feet (90 m) apart when the fruit will be picked by hand. For mechanically harvested fields, crossroads may be spaced farther apart, as long as access is provided to get equipment into the rows and to move irrigation lines, if necessary. Leave at least 25 feet (7.5 m) of headland at the ends of rows to turn equipment around. Tables 5.1 and 5.2 give the number of plants needed per acre or hectare.

Consider cross-pollination when laying out the planting. Because black currants are not always reliably self-fertile, cross-pollination is recommended to ensure consistent yields of large berries. Planting blocks of ten rows of one cultivar, followed by ten rows of another normally provides adequate cross-pollination. Another option is to plant one bush of a pollinizer about every 50 feet (15 m) within each row of the main cultivar. Plant pollinizers in a triangular pattern within adjacent rows to ensure that all bushes are within about 25 feet (7.5 m) of a pollinizer. These two planting methods are illustrated in

TABLE 5.1. Number of plants per acre.

| Distance between plants in rows (feet) | Distance between rows (feet) | | | | |
|---|---|---|---|---|---|
| | 4 | 6 | 8 | 10 | 12 |
| 2 | 5,445 | 3,630 | 2,722 | 2,178 | 1,815 |
| 3 | 3,630 | 2,420 | 1,815 | 1,452 | 1,210 |
| 4 | 2,722 | 1,815 | 1,361 | 1,089 | 907 |
| 5 | 2,178 | 1,452 | 1,089 | 871 | 726 |
| 6 | 1,815 | 1,210 | 907 | 726 | 605 |

*Note:* Figures assume that the entire area is planted and do not take into account roads or headlands.

TABLE 5.2. Number of plants per hectare.

| Distance between plants in rows (meters) | Distance between rows (meters) | | | | |
|---|---|---|---|---|---|
| | 1.2 | 1.8 | 2.4 | 3.0 | 3.7 |
| 0.6 | 13,889 | 9,259 | 6,944 | 5,556 | 4,505 |
| 0.9 | 9,259 | 6,172 | 4,630 | 3,704 | 3,003 |
| 1.2 | 6,944 | 4,630 | 3,472 | 2,778 | 2,252 |
| 1.5 | 5,556 | 3,704 | 2,778 | 2,222 | 1,802 |
| 1.8 | 4,630 | 3,086 | 2,315 | 1,852 | 1,502 |

*Note:* Figures assume that the entire area is planted and do not take into account roads or headlands.

Figure 5.1. Selection of pollinizing cultivars is discussed in Chapter 2.

As mentioned in Chapter 2, red currants, white currants, and gooseberries are generally considered self-fertile and set fruit without cross-pollination during most seasons. Research has shown, however, that yields and fruit sizes can be increased in these crops by using pollinizing cultivars. To ensure consistently high fruit set and

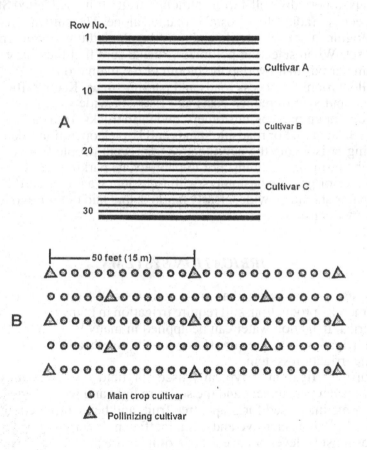

FIGURE 5.1. Planting designs that facilitate cross-pollination. (A) Alternating blocks of cultivars capable of pollinating each other. Each block consists of ten rows of a single cultivar. (B) Single bushes of a pollinizer every 50 feet (15 m) within rows. Stagger the locations of the pollinizers within adjacent rows.

quality in commercial operations, at least 1 to 2 percent of the plants in each field should be pollinizer cultivars. The pollinizer plants should be evenly distributed throughout the fields. Very little has been published on pollinizer selection for red and white currants and gooseberries, and most cultivars appear to be able to pollinate other cultivars of the same crop.

Jostaberries vary in their fertility according to cultivar. The cultivar Josta is, reportedly, self-fertile, although fruit set in the United States has been sporadic when 'Josta' is planted alone. The cultivars Jostine and Jogranda are self-sterile and must be planted together to ensure fruit set. When selecting pollinizers, make sure that they have long bloom periods that overlap the bloom of the main crop.

Allow room for access roads and parking areas. Keep traffic flow simple and safe to reduce costs and risks. Provide safe, convenient parking for employees, customers, and farm trucks from early spring through late fall. For roadside stands and U-pick operations, consider fencing or isolating the parking area to prevent people from driving into the farmyard or fields. For U-pick, locate parking areas close to the fields or provide customers with rides in a wagon or cart. Roads and paths should be wide, clearly marked, and laid out for maximum customer convenience.

## IRRIGATION SYSTEMS

*Ribes* use less water than many other small-fruit crops but, for commercial production, still require irrigation in many parts of North America. Irrigation water can be applied in many ways. Three common irrigation strategies are furrow or rill, overhead sprinklers, and trickle irrigation systems.

Furrow irrigation is typically used for field row crops in areas where water is abundant and the soils are light and well drained. It is also sometimes used for grape, tree fruit, and berry production. The method is labor intensive and often inefficient in terms of water use. Fields must be level or have a slope of no more than about 2 percent. Maintaining an even distribution of water throughout a field can be difficult.

Overhead systems typically used in small-fruit plantings include fixed and hand-line sprinklers and mobile water guns. If cover crops between berry rows are used, an overhead or microsprinkler system may be necessary. Overhead sprinklers can be adapted to almost any site and have many advantages. Sprinkling systems are

- relatively inexpensive to install,
- easy and inexpensive to maintain,
- provide generally uniform irrigation,
- can be used to protect fruit blossoms during spring frosts,
- can be used to cool foliage and fruit on hot days, and
- help to cleanse leaves and fruit of dust.

On the negative side, overhead irrigation systems waste water through evaporation and placing water between crop rows (which increases weed problems). Some sprinkler systems also require frequent movement of hand lines. Overhead irrigation wets foliage and fruit regularly and can increase disease problems.

Trickle or drip irrigation systems

- are efficient at placing water exactly where needed,
- use less water to produce a crop than sprinkler or furrow systems,
- reduce disease problems because the foliage and fruit are kept dry, and
- can be used to apply fertilizer to fruit crops.

Trickle systems, however, are often expensive to install and maintain and require filters and high-quality water to prevent emitters from clogging. Plugged drip emitters are harder to spot than plugged overhead sprinklers, and maintaining uniform irrigation requires constant scouting and maintenance. Trickle systems generally require level fields or must be laid across slopes to maintain even water pressure throughout each line.

In recent years, hybrid systems combining the best characteristics of sprinkler and trickle irrigation systems have been developed. These "microsprinklers" are easy to install and monitor but reduce the amount of water needed when compared with traditional overhead sprinklers. Properly designed and installed microsprinkler systems can provide frost control for the berries and irrigation for cover crops.

Where overhead sprinklers provide irrigation, about one acre-inch of water must be applied weekly to each acre during midsummer in dry regions. Where trickle systems are used and irrigation water is not required for a cover crop, only about one-half acre-inch of water is needed per acre, since only the crop row itself needs to be irrigated. An acre-inch of water equals 27,154 gallons (254,197 L/ha). On sites located in warm regions and/or on drought-susceptible soils, more irrigation may be needed. Tensiometers and gypsum or ceramic electrical resistance blocks can be used to help monitor soil moisture conditions in the root zone. These devices are readily available through farm supply stores.

Whatever irrigation system is used, the best practice is to have it designed by a professional irrigation specialist. Most firms that supply irrigation equipment can help with system design.


### DEER AND OTHER VERTEBRATE PEST CONTROL

Herbivore control can be an important part of planning and developing a berry farm. Moose and white tail, mule, and blacktail deer can be troublesome for fruit growers in some North American locations. As discussed earlier, while *Ribes* crops are susceptible to browse damage, they appear to be less attractive to herbivores than raspberries, blackberries, saskatoons, grapes, apples, pears, plums, and cherries. Moose and deer diets vary among regions, however, and *Ribes* growers would be advised to consult with local wildlife-management agencies on the need for herbivore control programs.

Many herbivore control methods have been tried, including repellant sprays, blood meal, coyote (or other predator) urine, soap bars tied to bushes, bags of human hair, and scarecrows. Unfortunately, herbivores are highly adaptable, and these methods seldom deter them from feeding on fruit plants for more than a few days to a few weeks. Also, some repellant products are not registered for fruit crops. If herbivore damage is a concern, the best method of control is to construct a specially designed fence.

Metal, chain-link fences effectively control deer and, to a lesser degree, moose but are expensive. The fence must extend to the ground at all points. Low spots under a fence allow deer to crawl through. Although deer are excellent jumpers, they will often crawl under or through a fence, rather than jump over it. Plastic mesh fences are an inexpensive alternative to metal mesh fences and are commercially available. The plastic mesh is not as durable as metal, of course, and can entangle an animal's feet. According to reports from growers, plastic fences can be seriously damaged when deer or moose become entangled in them. Keeping the fencing taught and staked tightly to the ground helps reduce entanglement problems. Where protection is needed only for several years while a crop becomes established, plastic mesh fencing appears to be an economical option.

A novel variation on mesh fencing for deer control was developed by a Pennsylvania apple grower, who constructed a short fence around his orchard. Although deer easily jumped over the fence into the orchard, the dogs the farmer kept there quickly chased them out. The dogs, a short-legged breed designed to herd reindeer, could neither jump the fence nor catch the deer. More recently, some commercial horticulturists have begun using buried electrical wires around their fields in combination with electronic collars on their dogs. As a dog approaches a buried wire, the collar emits a noise or harmless electrical shock. The dog soon learns where the boundaries are and generally remains in the fields. When chasing a deer or other animals, however, some dogs will ignore the collar warning and continue to give chase outside the fields. In some areas, free-roaming dogs or those that chase game animals can create legal problems for their owners.

Most fences designed to exclude herbivores are eight to ten feet (2.5 to 3 m) tall. Researchers in Pennsylvania have had some success with electrified, five-wire fences 58 inches (1.5 m) tall, but deer pressured by natural predators or human hunters will jump over the shorter fences. High-tensile steel wire should be used for electric fences. This system is often referred to as a New Zealand or Australian fence and is more durable and resilient than fences with low-tensile wire.

An effective and relatively inexpensive electrified fence is shown in Figure 5.2. For this design, place the bottom wire no more than six inches (15 cm) above the ground to prevent deer from crawling under the fence. Place additional wires at ten-inch (25 cm) intervals to a height of five feet (1.5 m), and at twelve- to sixteen-inch (30 to 40 cm) intervals above five feet. Electrify the bottom wire at all times. Electrify every other wire, so that when deer try to crawl between the

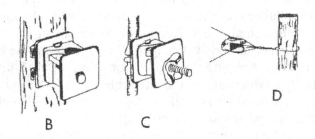

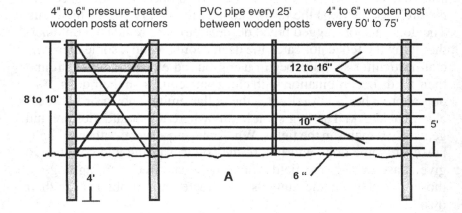

FIGURE 5.2. Diagram of an electrified herbivore fence (A). Note the wooden cross brace between corner posts. Also shown are electrical insulators for (B) wooden posts, (C) pipes, and (D) corner posts.

wires, they will contact at least one "hot" wire. Some growers recommend periodically changing which wires are electrified, always keeping the bottom wire electrified. The fence must completely enclose the field. Provide gates on all roads leading into the field.

Although generally effective in excluding herbivores, electrified fences are not infallible. Deer, elk, and moose have all been known to force their ways through fences, tolerating the shocks in order to feed. Tall grass and deep snow can short out wires. Regular inspections and repair are required to maintain a fence's effectiveness.

A word of caution: Electrified fences can be attractive to children and may constitute "attractive nuisances" if you live in or adjacent to a residential area. You may be liable if someone is hurt by your fence, even if that person does not have permission to be on the property. For farm sites in or near residential areas, mesh fences may be better than electrified systems.

## EQUIPMENT AND BUILDINGS

Most currant and gooseberry farms in North America are small and can be run with little specialized equipment. One of the most important items is a tractor. For very small operations, a heavy-duty garden tractor of approximately 15 horsepower may suffice, particularly for mowing cover crops. For all but the smallest operations, however, a small tractor in the 15 to 30 horsepower range and a width of about four feet (1.2 m) works well. Popular accessories include the following:

- Flail or deck mower for mowing cover crops and weeds
- Rototiller or other cultivator to control weeds
- Wagon or cart for hauling fruit, prunings, and equipment
- Backpack or tractor-mounted sprayer for applying pesticides

Currants, gooseberries, and jostaberries are usually picked by hand on small farms, but mechanical harvesters for processing fruit have been used for several decades in Europe. These machines range from handheld electrical vibrators to self-propelled over-the-row shakers. Mechanical harvesting is discussed in detail in Chapter 11.

Equipment storage areas should, preferably, be inside a building to reduce weathering, vandalism, and the chance of visitors being hurt on the equipment. Fuel storage tanks may also be needed. Ensure that the fuel storage meets government regulations. The same caution applies to pesticide storage. All pesticides should be locked up away from customer areas and areas where employees take their breaks.

Except for strictly U-pick operations, a refrigerated and/or controlled atmosphere room may be needed to store harvested fruit. Currants, gooseberries, and jostaberries are perishable, and their shelf life can be greatly increased by cooling them quickly after harvest. To preserve shelf life, cool the fruit to about 34°F (1°C) within two hours of picking. As described in Chapter 11, *Ribes* fruits are cooled dry by forcing cold air around and through the picking flats and containers, never by immersing the berries in or spraying them with water.

Cooling requires a refrigerator with adequate cooling capacity and a system of fans to circulate cold air through the flats of berries. Some growers have used refrigerated shipping trucks as fruit coolers. Although the trucks are adequate for keeping cold produce cold, they frequently lack the cooling capacity to remove field heat from the berries quickly. Prebuilt modular cooling units are available commercially or can be built on site. The capacity of the refrigeration system will depend on the amount of fruit to be cooled, the size of the room, and the insulation. Tables 5.3 and 5.4 give the amount of heat produced daily by respiration of stored gooseberry and black currant fruits. *Ribes* crops require approximately 900 cubic feet of refrigerated storage space per acre, although this rule of thumb assumes that the entire crop will be stored at one time, which may not always be the case. Allow additional space for storage racks, fans, and refrigeration equipment.

As discussed in Chapter 11, modifying the atmosphere around harvested berries can greatly increase the time the fruit can be stored and still retain acceptable quality. Although of little importance for processing fruit, which can easily be frozen, extending shelf life can greatly benefit fresh-market crops. In red currants, for example, mod-

TABLE 5.3. Respiration rates for gooseberry fruits and heat produced by fruit respiration.

| Temperature | | Respiration rate mg $CO_2 \cdot kg^{-1} \cdot h^{-1}$ | Heat produced | |
|---|---|---|---|---|
| °F | °C | | BTU per ton per day | kcal per tonne per day |
| 32 | 0 | 6-8 | 1,320-1,760 | 366-488 |
| 39-41 | 4-5 | 8-16 | 1,760-3,520 | 488-976 |
| 50 | 10 | 12-34 | 2,640-7,480 | 732-2,074 |
| 59-61 | 15-16 | 30-74 | 6,600-16,280 | 1,830-4,514 |
| 68-70 | 20-21 | 46-116 | 10,120-25,520 | 2,806-7,076 |

*Source:* Data adapted from Hardenburg and colleagues (1986), Robinson and colleagues (1975), and Smith (1967).

TABLE 5.4. Respiration rates for black currant fruits and heat produced by fruit respiration.

| Temperature | | Heat produced | | |
| --- | --- | --- | --- | --- |
| °F | °C | Respiration rate mg $CO_2 \cdot kg^{-1} \cdot h^{-1}$ | BTU per ton per day | kcal per tonne per day |
| 32 | 0 | 16 | 3,520 | 976 |
| 39-41 | 4-5 | 28 | 6,160 | 1,708 |
| 50 | 10 | 42 | 9,240 | 2,562 |
| 59-61 | 15-16 | 96 | 21,120 | 5,856 |
| 68-70 | 20-21 | 142 | 31,240 | 8,662 |

*Source:* Data adapted from Robinson and colleagues (1975).

ifying the atmosphere of refrigerated fruit can increase shelf life from around two weeks to more than 20 weeks, giving growers and brokers much more flexibility in marketing. Black currant shelf life can be doubled to approximately three weeks, while gooseberry shelf life can be increased from three weeks to as many as eight weeks.

Controlled atmosphere (CA) storage of currant and gooseberry fruits involves keeping the temperature near freezing, increasing carbon dioxide concentrations, and decreasing oxygen concentrations. Decreasing the temperature slows respiration and ripening while also retarding soft rots and other pathogens. Decreasing the oxygen concentration likewise slows respiration. Increasing carbon dioxide retards pathogen activity.

Controlled atmosphere systems have been used for many years for apples and other tree fruits, and have recently become popular for blueberries and other small fruits. What type of CA system to use will depend on your budget, amount of crop, and shelf life needed. CA storage facilities for apples usually involve modifying the atmosphere within entire rooms by using $CO_2$ generators, $CO_2$ scrubbers, oxygen scrubbers, and/or ethylene scrubbers. Such systems are very expensive and create risks for workers, as entering a CA room that has low oxygen and high carbon dioxide concentrations can easily be fatal. In contrast, some commercial blueberry packing houses successfully create CA storage environments by wrapping pallets of berries with selectively permeable plastic films and storing the pallets in conventional refrigerated rooms. Prior to storage, the berries have been cleaned, sorted, and placed into labeled containers and flats for consumers. Inside the plastic film, respiration from the berries uses up oxygen and increases carbon dioxide concentrations to CA standards. Careful monitoring of temperatures, relative humidity, and oxygen and carbon dioxide concentrations is important with both designs, but the latter system is much less expensive and greatly reduces risks associated with CA rooms. Prior to buying or building refrigeration or CA storage facilities, it would be advisable to visit several commercial fruit-packing houses and contact several equipment manufacturers and suppliers to determine which system best meets your needs.

# Chapter 6

# Preparing a Site for Planting

Proper site preparation begins at least one year in advance of planting. The goals are to improve drainage, correct soil nutrient deficiencies, adjust soil pH, eliminate perennial weeds, and reduce pest and pathogen problems. If soil drainage and aeration are concerns, one of the first tasks will be to improve the drainage by installing tiles or drainage ditches, or by leveling the site. Many small-fruit experts in the 1800s and early 1900s strongly recommended subsoiling with a chisel plow or other ripping device before planting currants and gooseberries (Sears, 1925).

## *AMENDING THE SOIL*

According to research, an important part of soil preparation for *Ribes* is the addition of large amounts of organic material to the soil. One Canadian publication recommended applying approximately 22 tons of farmyard manure per acre (49.5 tonnes/ha) before planting (Hughes, 1972). Sawdust or bark (preferably well rotted and composted) may also be used as soil amendments and mulches, although the amounts needed for large plantings can be prohibitive in terms of purchase and transport costs. Sawdust and bark may also be contaminated with *Armillaria,* a fungus that can kill currants, gooseberries, and jostaberries. Approximate volumes and coverage areas for sawdust and bark are listed in Table 6.1.

At a minimum, a green manure crop should be grown and tilled under the summer before planting. This practice adds organic matter to the soil, improves tilth, and helps control weeds. *Ribes* do not perform well when they are planted in a field immediately following a heavy sod crop or well-established alfalfa. Buckwheat, barley, clover, vetch, oats, beans, peas, and rape are excellent crops to precede a

TABLE 6.1. Approximate volumes and coverage areas for bark and sawdust used as soil amendments or mulches.

| Quantity | Large bark chips | Small bark chips | Sawdust |
|---|---|---|---|
| 1 ft$^3$ weighs | 15 lb | 14 lb | 12 lb |
| 1 yd$^3$ weighs | 405 lb | 378 lb | 324 lb |
| 1-inch layer covering 1 acre weighs | 27-28 tons | 25-26 tons | 21-22 tons |
| 1 yd$^3$ makes a 1-inch layer that covers | 323 ft$^2$ | 323 ft$^2$ | 323 ft$^2$ |
| Amount needed to cover 100 ft$^2$ 1 inch deep | 8.4 ft$^3$ | 8.4 ft$^3$ | 8.4 ft$^3$ |
| Amount needed to cover 1 acre 1 inch deep | 135 yd$^3$ | 135 yd$^3$ | 135 yd$^3$ |
| 1 m$^3$ weighs | 239 kg | 223 kg | 191 kg |
| 1-cm layer covering 1 hectare weighs | 23.9 tonnes | 22.3 tonnes | 19.1 tonnes |
| 1 m$^3$ makes a 1-cm layer that covers | 100 m$^2$ | 100 m$^2$ | 100 m$^2$ |
| Amount needed to cover 10 m$^2$ 1 cm deep | 0.1 m$^3$ | 0.1 m$^3$ | l0.1 m$^3$ |
| Amount needed to cover 1 hectare 1 cm deep | 100 m$^3$ | 100 m$^3$ | 100 m$^3$ |

*Note:* Figures are based on measurements of mixed conifer bark and sawdust. Materials from other tree species may differ in weight.

gooseberry or currant planting. Although verticillium root rot is not reported to be a severe problem in *Ribes,* it may be advisable not to plant gooseberries, currants, or jostaberries in a field immediately following potatoes, tomatoes, eggplants, or other small fruits.

If uncomposted woody organic materials (such as straw, sawdust, or wood chips) are incorporated as amendments into the soil, apply nitrogen to facilitate microbial breakdown of the amendment. Soil microorganisms use nitrogen as they break down organic materials. Unless extra nitrogen is applied with uncomposted, woody amendments, soil nitrogen can become temporarily depleted, creating nitrogen deficiencies in crops. When hardwood sawdust or bark is incorporated into the soil as amendments, apply 25 pounds of actual nitrogen for each ton (12.5 kg/tonne) of woody material. Apply 12 pounds of nitrogen per ton (6 kg/tonne) of softwood (conifer) material. The following year, add about half that amount of nitrogen to the respective fields, in addition to the amount of fertilizer needed for the crop.

To determine how much of a particular fertilizer to apply, divide the amount of nitrogen needed by the percentage of nitrogen in the fertilizer. Every fertilizer container lists the percentages of nitrogen, phosphorus, and potassium in that product. The percentage of nitrogen is always the first of the three numbers, followed by measures of phosphorus and potassium, respectively. A fertilizer designated as 10-5-15 has 10 percent nitrogen, 5 percent $P_2O_5$, and 15 percent $K_2O$. In this case, divide the amount of fertilizer needed (25 pounds) by the percentage of nitrogen (0.10) to determine that 250 pounds of 10-5-15 fertilizer are needed for each ton of uncomposted, hardwood sawdust or bark incorporated into the soil.

When applying fertilizer to break down woody organic materials, nitrogen is the most important nutrient. Unless soil tests show phosphorus and potassium to be deficient, it is more cost-effective to apply nitrogen fertilizers only, such as urea (46-0-0) or ammonium sulfate (21-0-0).

Composting organic materials before applying them to fields is highly recommended to control weeds, insects, and diseases. Where large amounts of manure are applied, test the soil yearly for salt concentrations and pH.

*Ribes* grow best on a neutral to slightly acidic soil, with optimum pH values between 5.5 and 7.0. On soils with pH values above 7.0 and concentrations greater than 15 percent calcium, iron chlorosis may become a problem. If the pH is marginally high, agricultural sulfur can be applied to an otherwise acceptable site one year before planting to lower soil pH, provided that the soil is not excessively alkaline. Sulfate fertilizers, such as ammonium sulfate [$(NH_4)_2SO_4$], will acidify the soil over a period of time. Acidifying fertilizers can be used to control soil pH on sites that are susceptible to alkaline conditions due to soil or water characteristics. If the soil pH is about 8.0 or above, large-scale acidification is usually both difficult and expensive. Limestone or dolomite applications can be used to raise the pH of soils that are too acidic. If liming materials are necessary, apply them the year before planting. The amount of sulfur or lime to add depends on the soil pH and exchange acidity, both of which are determined with a soil analysis. Follow the soil analytical lab's recommendations for the amount of lime or sulfur needed to adjust the soil pH. Finely ground materials react more quickly in the soil to alter pH than coarse materials.

In areas characterized by heavy precipitation and relatively pure irrigation water, soils tend to be acidic and even limed soils revert to an acidic condition within a few years. Fertilization practices also affect soil pH. For these reasons, soil pH should be determined with a soil analysis every two or three years. In fields where fertilizers are banded within crop rows (as opposed to broadcasting it throughout the field) and/or when trickle irrigation is used, soil pH can vary between the crop rows and alleys. In these cases, take separate soil samples both within and between rows and determine the pH for each area. Some growers may find it necessary to lime only within crop rows.

*Ribes* crops have the reputation of thriving on poor soils. This misconception probably arises from the fact that neglected currant and gooseberry bushes survive for long periods of time. To maintain healthy bushes and consistently high yields of quality fruit, however, proper fertilization is required. Nitrogen and potassium (potash) are the nutrients most often limiting in *Ribes* production.

Phosphorus ($P_2O_5$) and potassium ($K_2O$) soil analysis values of 100 and 150 ppm, respectively, generally indicate sufficient concentrations of these nutrients (Harmat et al., 1990). *Ribes* are reported to be sensitive to chlorine salts, and British authorities have recommended against the use of potassium chloride for currants and gooseberries (Shoemaker, 1948). Card (1907), however, reported that experiments in New York comparing potassium chloride and potassium sulfate showed no differences in the responses of currants to the two fertilizers. Table 6.2 gives the percentages of nutrients in some commonly used fertilizers.

Although nitrogen, phosphorus, and potassium are the nutrients most often deficient in cropping systems, ten other essential elements can be deficient in soil. Unfortunately, relatively little information is available on recommended soil nutrient levels for *Ribes* crops. The following recommendations have been developed for similar crops and provide a starting point for amending the soil. Once a plantation is established, foliar analyses give a more accurate picture of nutri-

TABLE 6.2. Approximate percentages of nutrients in various fertilizers.

| Material | Average percentage | | | | | | | | |
|---|---|---|---|---|---|---|---|---|---|
| | N | $P_2O_5$ | $K_2O$ | Ca | Mg | S | Mn | Fe | B |
| Ammonium nitrate | 34 | — | — | — | — | — | — | — | — |
| Ammonium sulfate | 21 | — | — | — | — | 24 | — | — | — |
| Bone meal (steamed) | 1 | 25-30 | — | 56 | — | — | — | — | — |
| Borated gypsum | — | — | — | 22 | — | 18 | — | — | 1 |
| Borax | — | — | — | — | — | — | — | — | 11 |
| Boric acid | — | — | — | — | — | — | — | — | 17 |
| Boron frits | — | — | — | — | — | — | — | — | 2-6 |
| Calcium sulfate | — | — | — | 22 | — | 17 | — | — | — |
| Calcium nitrate | 16 | — | — | 24 | — | — | — | — | — |
| Diammonium phosphate | 18 | 46 | — | — | — | 2 | — | — | — |
| Ferric sulfate | — | — | — | — | — | 9 | — | 16 | — |
| Ferrous sulfate | — | — | — | — | — | 12 | — | 20 | — |
| Iron chelate | — | — | — | — | — | — | — | 10 | — |
| Magnesium sulfate | — | — | — | — | 10 | 13 | — | — | — |
| Manganese sulfate | — | — | — | — | 19 | 33 | — | — | — |
| Manganese chelate | — | — | — | — | — | — | 12 | — | — |

83

TABLE 6.2 (continued)

| Material | N | P$_2$O$_5$ | K$_2$O | Ca | Mg | S | Mn | Fe | B |
|---|---|---|---|---|---|---|---|---|---|
| Manure | 10 | — | — | — | — | — | — | — | — |
| Potassium sulfate | — | — | 54 | — | — | 18 | — | — | — |
| Potassium chloride | — | — | 60 | — | — | — | — | — | — |
| Potassium nitrate | 13 | — | 44 | — | — | — | — | — | — |
| Potassium magnesium sulfate | — | 22 | — | 11 | 23 | — | — | — | — |
| Rock phosphate | — | 5 | — | — | — | — | — | — | — |
| Sodium pentaborate | — | — | — | — | — | — | — | — | 18 |
| Sodium tetraborate | — | — | — | — | — | — | — | — | 14 |
| Solubor | — | — | — | — | — | 12 | — | 20 | — |
| Superphosphate (20%) | — | 20 | — | 20 | — | — | — | — | — |
| Triple phosphate | — | 46 | — | — | — | 2 | — | — | — |
| Urea | 46 | — | — | — | — | — | — | — | — |

*Source:* Adapted from Pritts and Wilcox (1986).
*Note:* To convert between percent P and percent P$_2$O$_5$: % P · 2.29 = % P$_2$O$_5$ and % P$_2$O$_5$ · 0.44 = % P. To convert between percent K and percent K$_2$O: % K · 1.2 = % K$_2$O and % K$_2$O · 0.83 = % K.

tion in *Ribes* crops. Foliar analysis and fertilization practices are discussed in Chapter 9.

Soil concentrations of less than 30 ppm of soluble magnesium indicate that magnesium is deficient. Add 500 pounds per acre (565 kg/ha) of either magnesium sulfate (epsom salts) or potassium-magnesium sulfate (sul-po-mag). When soil pH is low, dolomitic limestone, which contains magnesium, can be used to offset magnesium deficiencies.

For proper *Ribes* growth and fruit development, calcium should be present in the soil at between 1,000 and 5,000 ppm. When calcium is deficient, add it to the soil as either lime (calcium carbonate) or gypsum (calcium sulfate). Lime will raise the soil pH, while gypsum has little effect on pH. When a soil test indicates less than 1,000 ppm of calcium, add 2,000 lb/acre (2,260 kg/ha) of actual calcium. For values between 1,000 and 5,000 ppm, add 1,000 lb/acre (1,130 kg/ha) of actual calcium. No additional calcium is needed when soil test values are greater than 5,000 ppm.

Soils should have at least 10 ppm of sulfur for optimum plant growth. When a soil analysis shows lower concentrations, add 30 lb/acre (34 kg/ha) of actual sulfur. Gypsum is an excellent source of both calcium and sulfur, and 175 lb/acre (200 kg/ha) will provide the needed sulfur. If boron is also deficient, borated gypsum can be used to supply calcium, sulfur, and boron at the same time.

Boron is needed by plants in very small amounts and can quickly become toxic when too much is applied. When soil test values show less than 0.5 ppm boron in the top 12 inches (30 cm) of soil, add 1 to 2 lb/acre (1 to 2 kg/ha) of actual boron. To avoid concentrating too much boron within crop rows and damaging the plants, broadcast soil-applied, boron-containing fertilizers throughout fields, rather than banding them within crop rows. Soluble boron fertilizers are applied directly to the foliage.

## CONTROLLING WEEDS

Weeds are among the most serious challenges berry growers face. The problem is particularly severe with *Ribes* because few herbicides are registered for use in these crops. The key is to get weeds under control before planting currants, gooseberries, or jostaberries. Annual weeds can be nuisances but are usually easy to control with cultivation, herbicides, cover crops, and other practices. Aggressive perennial weeds, such as quackgrass and some thistles, can become very serious in berry fields. Often these weeds spread by means of underground stems called rhizomes. Cultivating with a rototiller or by hand often increases the weed problem by chopping the rhizomes into small pieces, each of which then grows into a new plant. Grasses are especially difficult to eliminate once they become established in berry rows.

For weedy sites, strongly consider taking from one to three years to get weeds under control before planting *Ribes*. This practice is especially valuable for organic growers. During the one- to three-year period, weed control may include cultivation, fallowing, planting rotation and green manure crops, and/or herbicide applications. Where perennial weeds are a problem, translocatable herbicides, such as glyphosate, can be valuable. Some recommended rotation and manure crops are mentioned at the beginning of this chapter. A recommended program for weed control and other site preparation activities is given in Box 6.1.

Weed control also helps to control pests and diseases. For example, sedges are alternate hosts for the cluster cup rust that infects gooseberries. Native, escaped, and abandoned currant and gooseberries can serve as hosts for many pests and diseases and should be eradicated in and around fields before planting.

## CONTROLLING PESTS

Controlling weeds and wild *Ribes* help to reduce insect and mite problems in currant, gooseberry, and jostaberry crops. Vertebrate pests also create challenges for fruit farmers, with birds being among the most common and serious problems. Damage occurs immediately before and during harvest when birds feed on the ripe berries. A field preparation strategy that helps reduce later bird damage is to remove roosting sites. Birds often flock in large trees near berry fields. On some sites, squirrels can also become serious pests, as they feed on the berries. Eliminating trees and brush around a planting can help reduce bird and rodent problems.

Gophers are serious pests that are difficult to eliminate once they are established. An important step is to remove their food supply by

**BOX 6.1. Calendar of Activities for Preestablishment and Planting Years**

### Preestablishment Year

#### May to August

Grow a rotation or green manure crop, maintaining good weed control
Clean, cultivate, and fallow fields
Test soil for nutrients, pH, exchangeable acidity, organic matter, texture, and nematodes
Apply lime and sulfur, as indicated by soil tests
Conduct water percolation tests in fields
Test water supply for quality, flow rate, and capacity

#### September and October

Order plants for next fall
Design field layout and irrigation system
Order irrigation and fencing materials

### Planting Year

#### May

Apply fertilizers, as indicated by soil tests
Cultivate fields with discs, rototiller, or harrow
Install irrigation system, fences, and roads

#### June and July

Spray weeds with a translocatable herbicide (optional)
Cultivate
Plant a rotation or green manure crop
Spot spray or remove weeds by hand

#### August and September

Spot spray weeds with a translocatable herbicide (optional)
Plow or rototill to incorporate crop residue, then disc and pack fields
Deep rip with a chisel plow within planting rows, if needed
Lay out and mark planting rows

#### September and October

Plant bushes and irrigate immediately after planting
Mulch rows and establish rodent control program (optional)
Winterize irrigation system

cultivating and fallowing fields for one or more years before planting. This practice normally drives gophers into adjoining fields and windbreaks. Animals that remain can be easily detected by the presence of their burrow mounds. Trapping or baiting with poison grain can then be used to manage the remaining gopher population. Since gophers prefer to feed underground on plants with large, fleshy roots, avoid using these kinds of crops in rotation and green manure cropping.

# Chapter 7

# *Ribes* Cultivars

## CULTIVARS AVAILABLE IN NORTH AMERICA

No topic is more likely to create arguments between currant and gooseberry growers than which cultivars are best. Because of the avid (sometimes rabid) championing of one cultivar or another, no list of recommended cultivars will ever go unchallenged.

Because of previous restrictions on *Ribes* production and the wide variety of well-adapted native and introduced fruits, breeders in the United States and Canada conducted little work with currants and gooseberries after the early to mid-1900s. Popular European cultivars are often not available in North America or do not perform well here because of disease problems. Interest in *Ribes* culture is increasing in North America, however, and new cultivars are being introduced.

The naming of currants and gooseberries has been confused for centuries, and a given cultivar can be known by several different names. Although thousands of cultivars have been named, relatively few of them are commercially available in North America, and not all of those are well adapted to our growing conditions. The following sections and tables describe some cultivars that have performed well in North America or that show promise but are not yet fully tested.

With currants and gooseberries, proper cultivar selection is critical because a particular cultivar often performs very differently in different growing regions (Keeble and Rawes, 1948). For example, the red currant 'Cherry' performs well in Idaho. Growers in New York, however, have described its performance as mediocre. Test several cultivars at your site before deciding which ones to plant in large numbers. Other factors to consider are how the fruit is intended to be used and how it will be harvested. Some cultivars are well suited to both fresh and processing use, while others are more limited in their applications. Likewise, some cultivars respond well to mechanical harvesting, while others are difficult to harvest by machine or the canes and/or fruit are easily damaged in the harvesting process. Many good currant and gooseberry cultivars are available. Particularly for a commercial enterprise, choose those that provide consistent, high yields of quality fruit with a minimum of problems and expense. Important factors to consider when selecting cultivars include the following:

- Pest and disease resistance
- Fruit quality
- Yield
- Growth habit
- Suitability for mechanical harvesting

## DISEASE RESISTANCE

Except for powdery mildew and white pine blister rust, currants and gooseberries have the undeserved reputation of being relatively free from diseases. Many home gardeners and hobby growers in North America experience few problems with *Ribes* diseases because their plants are isolated from large populations of native or cultivated *Ribes*. As planting size increases, disease pressures do also.

The two most serious diseases of currants and gooseberries in North America are powdery mildew and white pine blister rust. Leaf spot diseases can also become serious problems in the northeastern United States and other warm, humid regions. These and other diseases and their control are discussed in detail in Chapter 10. Cultivars vary tremendously in their susceptibility to these diseases. While currants and gooseberries can host blister rust, for example, many red and white currants, gooseberries, and jostaberries suffer little damage from the disease. Black currants, on the other hand, are often highly susceptible to rust. Breeders have successfully developed black currants that are resistant and even immune to blister rust.

Powdery mildew can be extremely serious in *Ribes* plantings, depending on weather conditions. Currants and gooseberries are affected much more than commercially available jostaberries. A cool, wet spring followed by hot, dry or humid weather often results in severe mildew symptoms. Fruit can be rendered inedible, bushes defoliated, and new shoots killed by the disease. The disease is most serious on European gooseberries and black currants because both the foliage and fruit can be damaged. Powdery mildew may defoliate red and white currants, but the fruit is generally not affected.

At least three species of powdery mildew infect cultivated currants and gooseberries, one from Europe and two from North America. European powdery mildew *(Microsphaera grossulariae)* is not, apparently, found in North America. This disease is seldom serious on cultivars with European ancestry. American powdery mildew species *(Sphaerotheca mors-uvae* and *S. macularis)*, on the other hand, can devastate currants and gooseberries. European gooseberries are especially susceptible to American powdery mildew. According to Hedrick (1925, p. 311), "nowhere in the New World is the European gooseberry a commercial success." Thomas and Wood (1909, p. 393) observed for English gooseberries that

> many hundreds have been named and described, and large numbers have been imported and tried in [the United States], but they have so generally mildewed and become worthless after bearing a year or two that they have been mostly discarded.

According to Card (1907) a motion was made at the 1884 meeting of the American Pomological Society to strike all English gooseberry cultivars from their lists. The motion was barely defeated. American powdery mildew was introduced into Britain in approximately 1905 and has devastated the gooseberry industry in the United Kingdom. Keeble and Rawes (1948) considered the American form of powdery mildew to be the most serious disease facing British gooseberry growers. Because of the seriousness of the disease, gooseberries never regained their popularity as a commercial crop in Britain (Rake, 1958). Some black currant cultivars are also highly susceptible to powdery mildew, while symptoms on red and white currants are seldom serious.

Recently, horticultural-grade mineral oil (Hummer and Picton, 2001) and other fungicides have proven effective in controlling mildew on *Ribes*. Although this expands the range of prospective cultivars growers have to choose from, the authors believe that resistance to mildew remains an important consideration in selecting gooseberry and currant cultivars.

## GOOSEBERRY CULTIVARS

Although once an extremely popular fruit in England and other western European countries, gooseberries have largely fallen out of favor with commercial fruit growers. Two reasons for the decline in popularity are the problems associated with American powdery mildew and the labor-intensive nature of the crop. Most gooseberry cultivars are thorny, making them more difficult to prune, harvest, and generally care for than many other small fruits.

Selecting a gooseberry cultivar can be challenging because of the sheer numbers of cultivars developed over the centuries. Harmat and colleagues (1990) cited estimates that 3,004 red, 675 yellow, 925 green, and 280 white-fruited gooseberry cultivars had been named, making a total of 4,884 cultivars of gooseberries worldwide. Of these cultivars, relatively few have proven suitable for commercial use (Harmat et al., 1990).

Although many North American growers are interested in raising large-fruited European cultivars, the facts are that many European gooseberry cultivars are highly susceptible to American powdery mildew and have historically proven unsuitable for commercial production in North America. Chemicals are available to control powdery mildew on gooseberries, as described in Chapter 10. Planting disease-susceptible cultivars with the intent of combating diseases with fungicides and other compounds, however, increases management, labor, and chemical costs. Gooseberry cultivars that consistently perform well in North America, or which show some promise here, are described in the following sections. Table 7.1 describes the fruit quality and horticultural characteristics of selected gooseberry cultivars.

Russia, the third-leading European gooseberry producer, has recently favored the gooseberry cultivars Russkij, Szemena, Kohoznig, and Rekord (Harmat et al., 1990). Prior to devastation of the industry by powdery mildew, growers in the United Kingdom favored 'Careless', 'Lancashire Lad', 'Whinham's Industry', 'May Duke', and 'Keepsake' for processing, and 'Howard's Lancer', 'Whitesmith', and 'Leveller' for fresh market sales. Gooseberries popular in other European countries include: 'Weise Triumphbeere', 'Rote Triumphbeere', 'Weise Voltragenda', 'Lady Delamere', 'Hoennings Fruheste', 'Triumphant', and 'Green Giant' (Harmat et al., 1990).

TABLE 7.1. Gooseberry cultivar characteristics.

| Cultivar | Mildew resistance | Vigor | pH | Fruit TSS (%) | Weight (grams) | Bloom first | Bloom end | Ripening first | Ripening end | Picking ease |
|---|---|---|---|---|---|---|---|---|---|---|
| Achilles | 1 | 1 | 3.1 | 8.0 | 5.10 | 94 | — | 185 | 200 | 2 |
| Blood Hound | 1 | 1 | 3.2 | 17.5 | 6.63 | 97 | — | 183 | — | 1 |
| Canada 0273 | 1 | 2 | 2.9 | 9.0 | 3.94 | 100 | 122 | 183 | 200 | 3 |
| Captivator | 2 | 2 | 2.9 | 13.0 | 4.78 | 102 | 122 | 185 | — | 2 |
| Careless | 1 | 1 | 3.2 | 10.0 | 3.91 | 102 | — | — | — | — |
| Catherine | 1 | 1 | — | — | — | — | — | — | — | — |
| Colossal | 1 | 1 | 3.0 | 12.0 | 5.02 | 100 | — | — | — | 2 |
| Columbus | 1 | 1 | 3.1 | 11.5 | 5.58 | 102 | — | — | — | — |
| Crandall | 4 | 2 | — | — | — | — | — | — | — | — |
| D. Young | 1 | 1 | 3.1 | 12.0 | 3.85 | — | — | — | — | — |
| Downing | 1 | 1 | — | — | — | — | — | — | — | — |
| Fredonia | 1 | 3 | 3.3 | 10.0 | 4.50 | 100 | — | — | — | 2 |
| Friedl | 2 | 1 | 3.1 | 10.0 | 3.08 | 100 | 122 | 187 | 215 | 3 |
| Glenton Green | 1 | 3 | 2.9 | 15.5 | 1.21 | 95 | — | 187 | 200 | 3 |

TABLE 7.1 (continued)

| Cultivar | Mildew resistance | Vigor | pH | Fruit TSS (%) | Weight (grams) | Bloom first | Bloom end | Ripening first | Ripening end | Picking ease |
|---|---|---|---|---|---|---|---|---|---|---|
| Golda | 1 | 1 | 3.0 | 11.0 | 2.45 | 95 | — | — | — | 2 |
| Green Hansa | 2 | 1 | 2.9 | 12.5 | 3.60 | 95 | — | 190 | 215 | 2 |
| Hinnomaen Kelt. | 3 | 1 | 2.9 | 14.0 | 1.60 | — | — | — | — | — |
| Hinomakki | 1 | 1 | 3.1 | 11.5 | 3.38 | 93 | — | 183 | 200 | 2 |
| Hoennings Frue. | 1 | 1 | 3.2 | 14.0 | 3.93 | 96 | — | — | — | — |
| Houghton | 1 | 1 | — | — | — | — | — | — | — | — |
| Howard's Lancer | 1 | 1 | 3.2 | 14.0 | 3.50 | 96 | 114 | — | — | — |
| Invicta | 2 | 1 | 3.1 | 10.0 | 6.11 | — | — | — | — | — |
| Josselyn | 1 | 1 | 3.0 | 14.0 | 1.93 | 96 | — | — | — | — |
| Jubilee Careless | 1 | 1 | 3.3 | 9.5 | 1.93 | — | — | — | — | — |
| Jumbo | 1 | 1 | 3.3 | 15.0 | 7.37 | 100 | — | 187 | — | 2 |
| Lancashire Lad | 1 | 1 | 3.2 | 15.5 | 5.26 | 96 | 120 | 187 | 200 | 2 |
| Lepaa Red | 3 | 1 | 3.1 | 19.0 | 1.17 | — | — | — | — | 2 |
| Leveller | 1 | 1 | — | — | — | 98 | — | — | — | 2 |

| | | | | | | | | | |
|---|---|---|---|---|---|---|---|---|---|
| Mary Tennis | 1 | 3.0 | 13.5 | 3.65 | 100 | 122 | 192 | 215 | 3 |
| Oregon | 2 | 3.1 | 15.0 | 2.85 | 96 | — | 187 | 208 | 2 |
| O.T. 126 | 3 | 2.9 | 7.5 | 3.96 | 104 | 122 | 187 | 215 | 2 |
| Pixwell | 3 | 3.0 | 15.0 | 1.93 | 104 | 118 | 187 | 202 | 2 |
| Poorman | 3 | 2.9 | 18.0 | 1.46 | 100 | — | 187 | 207 | 2 |
| Rosko | 1 | 3.2 | 10.5 | 3.29 | 97 | — | — | — | — |
| Ross | 1 | 2.9 | 10.5 | 2.99 | 100 | 118 | 185 | 196 | 2 |
| Snowdrop | 1 | 2.9 | 17.0 | 2.80 | 100 | 118 | 187 | 200 | 2 |
| Speedwell | 3 | 3.0 | 14.0 | 3.21 | 96 | — | 187 | 207 | 2 |
| Spinefree | 1 | — | — | — | 104 | 125 | 187 | 201 | 3 |
| Surprise | 1 | 3.3 | 18.0 | 3.67 | 100 | — | 187 | 194 | 1 |
| Whitesmith | 3 | — | — | — | 100 | 116 | 186 | 207 | 2 |
| Worchesterberry | 2 | 2.7 | 11.0 | 1.84 | 96 | — | 183 | 208 | 2 |

*Note:* Mildew (American powdery mildew) resistance: 1 = poor, 2 = moderate, 3 = good, 4 = no symptoms; vigor: 1 = poor, 2 = moderate, 3 = good; picking ease: 1 = difficult, 2 = average, 3 = easy; bloom and ripening dates: 1 = January 1, 365 = December 31; data missing or not available: —.

In North America, 'Houghton' and 'Downing' were among the first popular American gooseberry cultivars (Hedrick, 1925) and are still grown today. Other popular gooseberries in North America include 'Poorman', 'Pixwell', 'Silvia' (or 'Sylvia'), 'Captivator', 'Oregon' (or 'Oregon Champion'), 'Welcome', 'Ross', and 'Red Jacket' (also known as 'Josselyn'). In research trials, the authors have noted moderate to severe powdery mildew on 'Red Jacket'.

In terms of European gooseberries suitable for North America, Harmat and colleagues (1990) recommended 'Chataqua', 'Clark', and 'Industry' (or 'Whinham's Industry'), which have shown some resistance to mildew. Hedrick (1925) considered 'Chataqua' the "most promising" European gooseberry for North America and believed that 'Keepsake' also had some promise. Hedrick also recommended 'Lancashire Lad' because of its relative resistance to mildew, yet the cultivar suffers severe powdery mildew symptoms in the Pacific Northwest. Although the berries are small, 'Lepaa Red' is a promising European cultivar. Other European gooseberries with some degree of potential for commercial culture in North America are 'Green Hansa', 'Invicta', 'May Duke', 'Whitesmith', and 'Worchesterberry'.

Although gooseberries destined for the dessert market must be harvested by hand, processing fruit can be machine harvested to reduce labor costs. Salamon and Chlebowska (1993) estimated that a picker can hand harvest from 13 to 35 pounds (6 to 16 kg) of gooseberries per hour, while mechanical harvesting increases that figure to 1,100 pounds (500 kg) per man-hour. When selecting cultivars for mechanical harvest, thornlessness and large fruit size are important characteristics (Salamon and Chlebowska, 1993). Thorns on the bushes puncture the fruits during the shaking process, and large berries appear to separate more easily from the canes.

The following descriptions of and recommendations for gooseberries and currants are based on *An Evaluation of the* Ribes *Collection at the National Clonal Germplasm Repository* (Barney and Gerton, 1992), *Small Fruits of New York* (Hedrick, 1925), germplasm trials by the authors in Idaho and Oregon, cultivar release notes from breeders, consultation with other researchers, and information in the U.S. Department of Agriculture Agricultural Research Service National Plant Germplasm System GRIN database (USDA, 2003b).

## Recommended American Gooseberries

### Captivator

'Captivator' has medium-sized, tear-drop-shaped, smooth fruit, greenish-red to red when ripe, with good flavor. Canes are moderately vigorous, erect, and less spiny than many other cultivars. It is only slightly susceptible to powdery mildew and therefore one of the most promising gooseberries for commercial production in North America.

### Downing

Fruit from 'Downing' ripens in midseason and is smooth, green, tough skinned, and of very good quality. Size, however, is variable, ranging from small to medium, and the berries can be difficult to pick. Although the flavor is good, the berries begin to decay soon after maturing, making it difficult to market fully ripened fresh fruit. The canes are above average in size and vigor, dense, thorny, and quite resistant to powdery mildew. It was the most widely grown gooseberry in the United States during the early 1900s. Bred in 1855 by Charles Downing of Newburg, New York, it is a seedling of 'Houghton' and has some European ancestry. Hedrick (1925) considered 'Downing' to be an inferior gooseberry for commercial use, recommending the English cultivar Chataqua instead.

### Glenndale

'Glenndale' canes are semierect, very thorny, and vigorous, reportedly reaching eight feet (2.5 m) tall in trials in the northeastern United States. Moderately resistant to powdery mildew, its moderate-quality berries are small and dark red to purple. The cultivar reportedly produces high yields in the northeastern United States. Released in 1932 by the U.S. Department of Agriculture in Glenn Dale, Maryland [(*R. missouriense* × 'Red Warrington') × 'Triumph'] × 'Keepsake', 'Glenndale' reportedly tolerates bright sun and was bred for growers at extreme southern limit of gooseberry culture (California Rare Fruit Growers, Inc., 2002).

## Houghton

The fruit ripens during midseason and is very small, dull red, moderately firm, very sweet, and has a pleasing flavor. The canes are very large and vigorous, erect to spreading, dense, and only slightly susceptible to powdery mildew. One of the first named American cultivars, it was very popular in North America during the early 1900s. Bred in 1833 in Massachusetts. Hedrick (1925) considered 'Houghton' a poor commercial cultivar because the fruit is small, unattractive, and not popular with fruit processors.

## Jahn's Prairie

A selection of *R. oxyacanthoides* L. collected in 1984 by Dr. Otto L. Jahn of the U.S. Department of Agriculture from the Red Deer River Valley in Alberta, Canada, it was released in 1996 by Dr. Kim Hummer of the USDA-ARS National Clonal Germplasm Repository in Corvallis, Oregon. 'Jahn's Prairie' is a disease-resistant, high-quality dessert gooseberry. Fruit is large, globose, red-pink, and ripens from mid to late July, producing high yields. Its habit is generally upright with some sprawling branches. Canes are thickly bristled with nodal spines. Growers in the northeastern United States report that the plants are somewhat slow to mature and develop high yields but they are resistant to powdery mildew, leaf spot, white pine blister rust, stem botrytis, aphids, and saw-flies.

## Oregon

Also known as 'Oregon Champion', the fruit ripens during the midseason, is small, white to pale green, tart, and has a thin, tough skin. The berries are, perhaps, better suited to processing than fresh markets. The canes are large, vigorous, erect to spreading, and quite resistant to powdery mildew. It originated in 1860 in Oregon as a cross between 'Crown Bob' and 'Houghton'.

## Pixwell

'Pixwell' fruit is small to medium in size but larger than 'Poorman'. It is pale green, ripening to pinkish-red, and hangs in clusters below the canes, making it easier to pick than some other cultivars.

The canes are vigorous, erect to spreading, and very resistant to powdery mildew. The fruit can develop a bitter taste in overripe berries. It is one of the most reliable gooseberries in North America, although fruit size, flavor, and overall fruit quality are marginal for fresh markets. This cultivar is better suited to processing. It was developed by the North Dakota Experiment Station in 1932 as a cross between *Ribes missouriense* and 'Oregon Champion'.

## Poorman

The dull-red fruit of 'Poorman' ripens over a long period beginning in midseason, is small to medium in size, sweet, aromatic, and has a tough, smooth skin. The canes are large, vigorous, erect to spreading, possess relatively few spines, and are quite resistant to powdery mildew. The cultivar reportedly originated as a cross between 'Houghton' and 'Downing' in Utah in 1888 (Shoemaker, 1948). Hedrick (1925) considered 'Poorman' to be one of the best gooseberries for North America, where it remains popular today.

## Sabine

'Sabine' canes are vigorous, erect to somewhat spreading, moderately spiny, and moderately susceptible to powdery mildew. The fruit size varies, typically being medium sized. Berries are pinkinsh in color, and yields are relatively low, but fruit quality is good. It was developed at the Central Experiment Farm in Ontario, Canada, by Blair and Hunter and introduced in 1950 as the result of the cross 'Spine-free' × 'Clark'.

## Sebastian

The vigorous, erect canes of 'Sebastian' have few spines and are moderately susceptible to powdery mildew. The medium-sized fruits are reddish in color and reported to have good to very good quality. Yields are high. It was developed at the Central Experiment Farm, Ontario, Canada, by A. W. S. Hunter.

*Shefford*

In 'Shefford' berry flavor and size are good, although yields are, reportedly, low. It was developed at the Central Experiment Farm, Ontario, Canada, by A. W. S. Hunter.

*Silvia*

Also known as 'Sylvia', the fruit ripens in midseason and is green tinged with dull red. Berries are large, subacid, and have good to very good flavor. Yields are low. The canes are strong, erect to spreading, and relatively free of powdery mildew. It was developed in Canada in the late 1890s.

*Stanbridge*

'Stanbridge' canes are erect, vigorous, and develop few or no spines. Resistant to powdery mildew, fruits are small and yellowish-green and yields are moderate. It was developed at the Central Experiment Farm, Ontario, Canada, by A. W. S. Hunter.

*Sutton*

'Sutton' was developed at the Central Experiment Farm, Ontario, Canada, by A. W. S. Hunter.

*Welcome*

'Welcome' originated as an open-pollinated seedling of 'Poorman' and was introduced by the Minnesota Agricultural Experiment Station in 1957. It is similar to 'Pixwell', but the fruit is larger, more uniform, and ripens earlier. The canes are relatively free of spines and quite resistant to powdery mildew and anthracnose. The fruit is a light, dull red and has high quality. Yields are variable, but it is widely available in the United States.

## Recommended European Gooseberries

### Chataqua

Sometimes confused with 'Whitesmith', the origin of 'Chataqua' is unknown. First described in the United States in 1876, it is probably an English cultivar that was renamed. The fruit ripens during the midseason and is large, smooth, green, tough skinned, flavorful, and has good quality. The canes are dense, spreading, and moderately erect. Moderately resistant to powdery mildew, Hedrick (1925) considered 'Chataqua' to be the "most promising" European gooseberry for North America.

### Clark

The fruit of 'Clark' is large and red. The canes are vigorous and spreading. Moderately susceptible to powdery mildew, the pedigree is, apparently, unknown. A chance seedling discovered by M. C. Smith of Ontario, Canada, was released as a cultivar in 1922.

### Green Hansa

'Green Hansa' green fruits are medium to large, green, and suitable for dessert use. The canes are below average in size and vigor and moderately resistant to powdery mildew. It is from Germany.

### Industry

Also known as 'Whinham's Industry', the red fruit ripens early in the season, ranges from medium to large in size, has a smooth to slightly hairy, moderately tough skin, and is of very good quality. The canes are vigorous and large for a European cultivar. Moderately susceptible to powdery mildew and difficult to propagate, the cultivar originated in England and was introduced into North America in about 1855. It was one of the most popular European gooseberries in North America during the late 1800s and early 1900s. Reportedly, 'Industry' remains a popular commercial cultivar in Holland.

## Invicta

The yellow fruits of 'Invicta' are very large, early ripening, tart, and green. Flavor ratings have been marginal in some trials and fruit quality has been reported to be variable. Very high yielding, the canes are below average in size and vigor, spiny, and moderately resistant to powdery mildew and white pine blister rust. The foliage and fruit are, reportedly, susceptible to leaf spot. Other reports indicate particular susceptibility to imported currant worm. This cultivar has been used for commercial-grower and U-pick fresh and processing markets. Released in 1981 by the Horticulture Research Institute, East Malling, Maidstone, Kent, England, United Kingdom, it is the product of the cross ['Resistenta' × 'Whinham's Industry'] × 'Keepsake'.

## Keepsake

'Keepsake' fruit is green with a red blush near the end, variable in size but tending to be small to medium, with a thin, tough, fairly smooth skin. Fruit quality is good, but the variability in berry size limits this cultivar's commercial value. The canes are moderately vigorous, medium sized, and reported to be fairly free from mildew. It was developed in the early 1800s in England.

## Lepaa Red

The fruit of 'Lepaa Red' is small, red, and has good flavor. The canes are small, vigorous, dense, productive, and highly resistant to powdery mildew and other diseases. One of the most reliable gooseberries in Idaho and Oregon trials, it has performed well in the northeastern United States as well. Although suited for processing because of the small berry size, the desirable flavor of the fresh fruit may make this cultivar a good choice for farmers' markets. It was developed in Finland.

## May Duke

'May Duke' fruit ripens early in the season, is medium to large in size, green becoming dull red when ripe, pleasantly flavored, and has a smooth, moderately tough skin. The canes are medium in size, vigorous, dense, erect, and moderately resistant to powdery mildew. It was brought to America in the early 1900s.

## Speedwell

The fruit of 'Speedwell' is large, nearly smooth, red, sweet, and has fair quality. The canes are moderately vigorous and relatively resistant to powdery mildew. The cultivar originated in England in the early 1800s.

## Whitesmith

Sometimes confused with 'Chataqua', the fruit ripens early in the season, is above average in size, green to light yellow, with a smooth, tender skin and sweet, pleasant flavor. The canes are average to above average in size, erect to spreading, and productive. Unfortunately, they tend to mildew rather badly. 'Whitesmith' is included in this list because in the early 1900s some growers considered it "to be the best of the English cultivars grown on this side of the Atlantic" (Hedrick, 1925, p. 353). Probably developed in England in the late 1700s, it is reportedly enjoying a resurgence of popularity in England and Scotland.

## BLACK CURRANT CULTIVARS

Black currants are widely grown in Europe, although spring frosts have historically caused frequent damage and highly variable yields in some regions. 'Baldwin' formerly made up 80 percent of the acreage in Britain (Harmat et al., 1990). Although possessing excellent flavor, 'Baldwin' is highly susceptible to spring frosts. Introduction of high northern latitude germplasm has allowed the development of more widely adapted cultivars, such as 'Ben Lomond' and 'Ben Nevis' from the Scottish Crop Research Institute. These cultivars are more resistant to frost damage but retain the fruit qualities of southern genotypes. In 1997, the Scottish Crop Research Institute released 'Ben Hope' and 'Ben Gairn', based on their resistance to gall mite and black currant reversion disease, respectively. Despite these new introductions, many older, locally derived cultivars are grown (Brennan, 1990). 'Ojebyn' remains widely popular in Scandinavia, Finland, and Poland, with localized production of 'Melalahti', 'Matka-kowski', and 'Roodknop'. Increasing in popularity are 'Titania', 'Stor Klas', and some of the Scottish 'Ben' cultivars (Brennan, 1996). Countries in the former Soviet Union grow many diverse black currant cultivars, including the highly-productive 'Golubka', 'Narjad-naja', 'Brodtorp', 'Vystavochnaja', and 'Stakhanovka Altaja'. Less productive, but still popular, cultivars include 'Belorusskaja Slodkaja', 'Minaj Smirev', and 'Altajskaja Desertnaja'. 'Tenah' and 'Tsema' are traditional favorites in Holland. 'Noir de Bourgogne' is still grown in France but may be replaced by the newer cultivars 'Tifon' and 'Troll' (Brennan, 1996). In New Zealand, northern European cultivars, including the 'Ben' series, are replacing 'Magnus', 'Blackdown', and 'Topsy'.

Keep disease resistance in mind when selecting black currants. If white, sugar, whitebark, bristlecone, or other five-needled pines grow in your region, grow blister rust-resistant black currant cultivars. The Canadian cultivars 'Consort,' 'Coronet,' and 'Crusader' are resistant (Keep, 1975) due to a dominant rust-resistant gene derived from *R. ussuriense* (Hunter, 1950, 1955). The flavor of these cultivars, however, does not meet the standards of many European cultivars and all are highly susceptible to powdery mildew. Serengovyj (1969) reported the European cultivar 'Primorskij Cempion' (*R. nigrum* × *R. dikuscha*) to be resistant to blister rust. Somorowski (1964) reported that *R. nigrum europeaum* × *R. nigrum sibiricum* hybrids 'Bzura', 'Dunajec', 'Loda', 'Ner', 'Odra', 'Warta', and 'Wista' showed field resistance to both rust and leafspot. The 1984 Swedish introduction 'Titania' has blister rust resistance genes from both *R. dikuscha* and *R. ussuriense*, and has generally been reported as blister rust resistant. Blister rust symptoms were observed, however, on 'Titania' in Danish trials. 'Titania' appears to be resistant to powdery mildew. Although its juice-processing quality is marginal by European standards, the flavor and overall quality of 'Titania' exceed those of 'Consort', 'Crusader', and 'Coronet'. In recent evaluations by the author, Dr. Kim Hummer, the following black currant cultivars were classified as immune or highly resistant to blister rust: 'Consort', 'Coronet', 'Crandall', 'Crusader', 'Doez Sibirjoczk', 'Lowes Auslese', 'Lunnaja', 'Polar', 'Rain-in-the-Face', 'Sligo', 'Titania', and 'Willoughby'. Note, however, that different strains of the blister rust pathogen are distributed within North America and that a cultivar resistant or immune to one race may be susceptible to another race (Hansen, 1979). With research lacking on black currant cultivars' rust resistance in different growing regions, growers should test selected cultivars in small plantings for several years before investing heavily in large-scale plantings.

Another consideration in selecting black currant cultivars is suitability for mechanical harvesting. In European trials, for example, 'Ben Nevis' and 'Ojebyn' responded well to mechanical harvesting, while canes of 'Roodknop' and 'Titania' were damaged by the shaking process (Salamon, 1993). The harvester design is also a consideration, however. In other trials (Tahvonen, 1979), 'Roodknop' exhibited less damage than did 'Ojebyn'. Table 7.2 gives the results of the authors' evaluations of black currants.

## Recommended Black Currants

### Ben Alder

The fruit is similar to 'Baldwin' in size and produces juice of very high quality. The fruit and juice exhibit high color stability. Late flowering makes this cultivar somewhat resistant to spring frosts. It is high yielding under ideal conditions, but low yields have been reported from the northeastern United States. Northeastern growers have also reported marked susceptibility to white pine blister rust. The canes are vigorous, can become dense, and show more mildew resistance than 'Ben Lomond' and 'Baldwin'. The cultivar yields better than 'Baldwin' and is seen as a replacement for it. The fruit can be harvested mechanically. It was developed by Malcolm Anderson at the Scottish Crop Research Institute. ('Ben Lomond' × 'Ben More').

### Ben Connan

'Ben Connan' berries are large, deep black, and considered suitable for commercial fresh and processing markets, as well as for home production, although not ideal for juice markets. It has good resistance to American powdery mildew and leaf curling midge, and it reportedly tolerates spring frosts and has produced high yields in British tests. It shows a compact growth habit and is suitable for machine harvesting. It was released by the Mylnefield Research Services as a cross between 'Ben Sarek' and 'Ben Lomond'.

TABLE 7.2. Black currant cultivar characteristics.

| Cultivar | Mildew resistance | Vigor | pH | Fruit TSS (%) | Weight (grams) | Bloom | | Ripening | | Picking ease |
|---|---|---|---|---|---|---|---|---|---|---|
| | | | | | | first | end | first | end | |
| Alagan | 2 | 2 | 2.7 | 19.5 | 0.80 | 97 | 118 | 177 | 192 | 3 |
| Ben Sarek | 2 | 1 | 2.6 | 16.0 | 0.77 | — | — | — | — | — |
| Black Sept. | 3 | 3 | 2.9 | 17.5 | 1.04 | 100 | — | 183 | 208 | 3 |
| Bogatyr | 2 | 1 | 2.9 | 17.5 | 0.53 | — | — | — | — | — |
| Boskoop Giant | 2 | 3 | 2.8 | 18.0 | 0.79 | 96 | 120 | 183 | 192 | 3 |
| Champion | 3 | 2 | 2.8 | 19.5 | 0.67 | 96 | 125 | 183 | 215 | 2 |
| Consort | 2 | 2 | 2.7 | 15.0 | 0.58 | 100 | 114 | 175 | — | 2 |
| Coronet | 2 | 3 | 2.6 | 16.5 | 0.62 | 100 | 120 | 183 | 196 | 2 |
| Crusader | 2 | 3 | 2.7 | 18.0 | 0.61 | 100 | 122 | 183 | 196 | 2 |
| Invigo | 2 | 3 | 2.8 | 18.0 | 0.75 | 100 | 118 | 183 | 200 | 3 |
| Kerry | 2 | 3 | 2.8 | 17.0 | 0.77 | 96 | 124 | 184 | 194 | 3 |
| Malvern Cross | 2 | 1 | 3.0 | 19.5 | 0.56 | — | — | — | — | — |

| | | | | | | | | | | |
|---|---|---|---|---|---|---|---|---|---|---|
| Noir de Bourg. | 3 | 2 | 2.8 | 15.5 | 0.63 | 95 | 114 | — | — | 3 |
| Pinot Debourk. | 2 | 2 | 2.8 | 19.0 | 0.51 | — | — | — | — | — |
| Raven | 3 | 1 | 2.9 | 15.5 | 0.52 | — | — | — | — | — |
| Saunders | 2 | 1 | 2.9 | 22.5 | 0.38 | — | — | — | — | — |
| Seabrook's Black | 2 | 1 | 2.7 | 16.0 | 0.45 | — | — | — | — | — |
| Strata | 2 | 2 | 2.8 | 17.5 | 1.28 | 95 | 116 | 183 | 194 | 2 |
| Swedish Black | 2 | 2 | 2.9 | 26.0 | 0.86 | 98 | 114 | 177 | 190 | 2 |
| Tenah | 3 | 2 | 2.7 | 16.5 | 0.79 | — | — | — | — | — |
| Topsy | 2 | 3 | 2.9 | 19.0 | 0.72 | 96 | 118 | 183 | 200 | 2 |
| Tsema | 3 | 1 | 2.7 | 18.5 | 0.59 | — | — | — | — | — |
| Wellington XXX | 2 | 1 | 2.9 | 16.0 | 0.46 | — | — | — | — | — |

*Note*: Mildew (American powdery mildew) resistance: 1 = poor, 2 = moderate, 3 = good, 4 = no symptoms; vigor: 1 = poor, 2 = moderate, 3 = good; picking ease: 1 = difficult, 2 = average, 3 = easy; bloom and ripening dates: 1 = January 1, 365 = December 31; data missing or not available: —.

### Ben Lomond

The berries of 'Ben Lomond' are larger than 'Baldwin', have a tough skin, and ripen evenly on the strigs. It blooms later than 'Baldwin' and tolerates lower temperatures during bloom, making this cultivar a good choice for frosty sites. Yields are greater and more consistent than for 'Baldwin'. The canes are compact, spreading, and moderately vigorous. Moderately tolerant of powdery mildew but susceptible to white pine blister rust, it is recommended for commercial production for both fresh and processing markets. Suitable for machine harvesting, its large, attractive fruit also has potential for U-pick markets. It was developed by Malcolm Anderson at the Scottish Crop Research Institute and released in 1975 as a result of the cross ('Consort' × 'Magnus') × ('Brodtorp' × 'Janslunda').

### Ben Nevis

'Ben Nevis' is generally similar to its sibling 'Ben Lomond', but the bushes are more spreading. In Northwest trials, it proved more susceptible to powdery mildew than 'Ben Lomond' and the flavor was rated poorer. It was bred by Malcolm Anderson and released from the Scottish Crop Research Institute in 1974 as the product of the cross ('Consort' × 'Magnus') × ('Brodtorp' × 'Janslunda').

### Ben Sarek

The berries of 'Ben Sarek' are large, have a tough skin, and ripen evenly. Clusters are short and close to the bud, making handpicking difficult. The fruit, however, can easily be shaken off using mechanical harvesters. Yields are high to very high under ideal conditions but have been relatively low in some trials in the northeastern United States. It is resistant to American powdery mildew and well suited to home production, but not recommended by the breeders for commercial juice production. A sibling of 'Ben More' and the first in a line of small-bush hybrids designed for high-density plantings, it was developed by Malcolm Anderson at the Scottish Crop Research Institute as an open pollinated seedling of 'Goliath' × 'Ojebyn'.

## Boskoop Giant

'Boskoop Giant' berries are large and ripen during the early to mid-season. The clusters are long and loose and hang on the bush well after ripening. The canes are large, vigorous, and moderately resistant to powdery mildew. Often used as a pollinizer of 'Silvergieters Zwarte' in Europe, it was introduced from Holland via England between 1895 and 1900 and was very popular for commercial production in the United States during the early 1900s. Nursery stock labeled as 'Boskoop Giant' is not always true to name. When pollination is poor, berries near the tips of the clusters tend to "run off" or abort after they appear to be set (Keeble and Rawes, 1948). The fragrant leaves and buds are, reportedly, well suited for use in teas.

## Champion

The fruit of 'Champion' is variable in size (generally medium to large) and mild flavored with good to very good quality. The berries, however, do not ripen evenly. The canes are vigorous, productive, and moderately resistant to powdery mildew. One of the most popular black currants for commercial production in the United States during the early 1900s, it was developed, probably from a chance seedling, in England around the 1870s.

## Consort

The fruit flavor and overall quality in 'Consort' are poor to fair and berry size is small to medium. Developed by A. W. S. Hunter at the Central Experiment Farm in Ottawa, Canada as a cross between 'Kerry' × *R. ussuriense,* it was released in 1951 to replace 'Crusader' and 'Coronet'. The cultivar is resistant to white pine blister rust but highly susceptible to powdery mildew. It served as a parent of 'Ben Lomond'.

## Coronet

A sibling of 'Crusader', 'Coronet' was developed by A. W. S. Hunter at the Central Experiment Farm in Ottawa, Canada, as a cross between *R. ussuriense* × 'Kerry'. Released in 1948, its fruit quality and size are similar to 'Consort' and 'Crusader'. The cultivar is highly self-sterile and requires cross-pollination. It is resistant to white pine blister rust but highly susceptible to powdery mildew.

## Crusader

'Crusader' is a sibling of 'Coronet', developed by A. W. S. Hunter at the Central Experiment Farm in Ottawa, Canada, as a cross between *R. ussuriense* × 'Kerry'. Released in 1948, its fruit quality and size are similar to 'Consort' and 'Crusader'. It is highly self-sterile and requires cross-pollination. This cultivar is resistant to white pine blister rust but highly susceptible to powdery mildew.

## Kerry

The berries of 'Kerry' are medium to large in size and ripen during the mid to late season. The canes are vigorous, productive, and moderately resistant to powdery mildew. A seedling of 'Naples', it was introduced as a cultivar in 1907 by W. Saunders of the Canada Department of Agriculture.

## Strata

The berries of 'Strata' can be very large, with a sweet, good flavor suited for fresh use. The canes are moderately vigorous and moderately resistant to powdery mildew. Yields are, reportedly, low in trials in the northeastern United States. It is a selection of *R. nigrum* from Germany.

## Swedish Black

'Swedish Black' berries are large and very sweet, having the highest sugar content and one of the best flavors of any black currant evaluated at the University of Idaho. The canes are moderately vigorous and more resistant to powdery mildew than most other European black currants in northwest trials. Although a promising cultivar, the canes tend to sprawl, making management difficult. Yields are, reportedly, low. It was apparently a selection of *R. nigrum* from Sweden.

## Titania

Canes of 'Titania' are tall, vigorous, and high yielding but tend to sprawl under heavy crop loads. Highly self-fertile, its very large berries are produced on medium-long strigs and ripen during midseason. The fruit ripens uniformly and keeps well on the bush. Berries have high acidity, low flavor, and are resistant to runoff or premature abscission, although a bitter fruit flavor has been reported in some trials. It is resistant to American powdery mildew and, with the exception of one trial in Denmark, has been reported to be resistant to white pine blister rust. Although 'Titania' is susceptible to aphids, red spider mite, and leaf spot, it is suitable for commercial production due to its high yields and good processing quality, and for home gardens due to its disease resistance. It can suffer significant cane damage during mechanical harvesting. Developed by P. Tamas and introduced in Sweden in 1984 ['Golubka' × ('Consort' × 'Wellington XXX')], it is an excellent choice in areas where white pine blister rust is a concern.

## Topsy

The fruits of 'Topsy' are large, thick skinned, and ripen during the mid-season. The canes are vigorous and moderately resistant to powdery mildew. It was developed in Canada in approximately 1890, reportedly as a cross between 'Dempsey's Black Currant' and 'Houghton' gooseberry.

One black currant that can cause confusion is 'Crandall'. Although most domestic currants are largely derived from the European species *R. nigrum, R. dikuscha,* and/or *R. ussuriense,* this cultivar is a selection of the North American native *R. aureum* var. *villosum* D.C. (formerly *R. odoratum* Wendl.), also known as golden currant. This species is used most often as an ornamental and can reach heights of eight to ten feet (2.5 to 3 m), although 'Crandall' has never grown more than four feet (1.2 m) tall in Idaho or Pennsylvania trials. The canes are rather weak and droop to the ground under crop loads, making trellises desirable. During the spring and summer, the plants produce yellow flowers. Very large, black berries ripen during late summer to early fall. Although some consider the taste unpleasant (Hedrick, 1925), the authors have found the flavor of fully ripe fruit mild, pleasant, and very different from European black currants. 'Crandall' performs well in areas having hot summers and exhibits few or no symptoms of powdery mildew. In Oregon tests, 'Crandall' proved highly resistant or immune to white pine blister rust.

## RED CURRANT CULTIVARS

Red currants are easy to grow, and many good cultivars adapted to a wide range of sites are available. In northern Europe, 'Jonkheer van Tets' is the leading cultivar, with yields of 4.5 to 7 tons per acre (10 to 15.8 tonnes/ha). 'Rondom' is also popular but reportedly is suscepti- ble to mycoplasma diseases. Also gaining popularity in Europe are 'Rovada', 'Rosetta', and 'Rotet'. In southern France and other warm climate areas, 'Jonifer' and 'Fertodi hoszszfurtu' perform well. In England, 'Earliest of Fourlands' (also known as 'Erstling aus Vier- landen'), 'Jonkheer van Tets', 'Laxton No. 1', 'Laxton's Perfection', 'Minnesota No. 71', 'Red Lake', 'Rondom', 'Stanza', and 'Wilson's Long Bunch' (also known as 'Victoria') (Wood, 1964) are popular (Harmat et al., 1990). Russian growers value the hardy 'Red Dutch', while currant farmers in Finland prefer 'Nortun', 'Fortun', and 'Jon- tun', all of which were derived from crosses with 'Jonkheer van Tets'. 'Redstart' (developed in the United Kingdom) and other newer cul- tivars are gaining in popularity in Europe (Brennan, 1990).

When currant production was at its peak in the United States, Darrow (1919, p. 30) stated that 'Fay', 'Perfection', 'Cherry', 'Red Cross', and 'London' (also known as 'London Market') "have been found entirely hardy in North Dakota and should be hardy anywhere in the United States." Darrow further recommended 'Perfection', 'Wilder', and 'Red Cross' for the Northeast; 'London Market', 'Wilder', 'Red Cross', and 'Perfection' for Michigan and other parts of the Midwest; and 'Perfection', 'London Market', 'Red Cross', 'Wilder', and 'Fay' for the Pacific Coast.

Although generally less susceptible to blister rust than black cur- rants, many red cultivars can become infected with the disease. In areas where blister rust is of concern, consider planting blister rust– resistant cultivars. The following cultivars proved immune or highly resistant to blister rust in Oregon tests: 'London Market', 'New York 72', 'Rondom', and 'Viking'. Table 7.3 gives the result of the authors' evaluations of selected red currants.

TABLE 7.3. Red currant cultivar characteristics.

| Cultivar | Mildew resistance | Vigor | pH | Fruit TSS (%) | Weight (grams) | Bloom first | Bloom end | Ripening first | Ripening end | Picking ease |
|---|---|---|---|---|---|---|---|---|---|---|
| Bronze | 1 | 3 | 2.7 | 13.0 | 0.78 | 94 | 118 | — | — | 2 |
| Cascade | 2 | 1 | 2.8 | 13.0 | 0.96 | 96 | 124 | 183 | 193 | 2 |
| Cherry | 3 | 3 | 2.8 | 13.0 | 0.54 | 96 | — | 183 | 190 | 2 |
| Fay | 2 | 2 | 2.8 | 11.0 | 0.63 | 93 | — | 183 | 208 | 2 |
| Heros | 2 | 3 | 2.8 | 12.5 | 0.66 | 94 | 118 | 177 | 197 | 2 |
| Jonk. Van Tets | 2 | 2 | 2.7 | 11.0 | 0.72 | 100 | — | 183 | 193 | 2 |
| Laxton No. 1 | 2 | 2 | 2.8 | 9.5 | 0.92 | 91 | 118 | 187 | 194 | 3 |
| London Market | 2 | 1 | 2.7 | 12.0 | 0.42 | 100 | 122 | 183 | 215 | 2 |
| Malling Redstart | 1 | 1 | 2.7 | 12.0 | 0.64 | 91 | 118 | 187 | 215 | 2 |
| Minnesota 52 | 2 | 2 | 2.9 | 14.0 | 0.49 | 92 | 122 | 183 | 195 | 3 |
| Minnesota 69 | 2 | 2 | 2.7 | 13.0 | 0.70 | 91 | 118 | 177 | 190 | 2 |
| Minnesota 71 | 2 | 2 | 2.7 | 13.0 | 0.71 | 96 | 118 | 183 | 208 | 2 |
| Mulka | 3 | 2 | — | — | — | 102 | — | 183 | 208 | 2 |
| NY 53 | 2 | 1 | 2.9 | 11.5 | 0.48 | 93 | 116 | — | — | 2 |
| NY 68 | 2 | 2 | 2.7 | 11.0 | 0.79 | 91 | 114 | — | — | 2 |
| NY 72 | 3 | 2 | 2.6 | 10.5 | 0.55 | 95 | 122 | — | — | 2 |
| Perfection | 2 | 3 | 2.6 | 13.5 | 0.97 | 93 | 120 | 183 | 200 | 2 |
| Portal Ruby | 2 | 2 | 2.8 | 10.5 | 1.06 | — | — | — | — | — |

TABLE 7.3 (continued)

| Cultivar | Mildew resistance | Vigor | pH | Fruit TSS (%) | Weight (grams) | Bloom | | Ripening | | Picking ease |
|---|---|---|---|---|---|---|---|---|---|---|
| | | | | | | first | end | first | end | |
| Prince Albert | 3 | 2 | 2.8 | 12.5 | 0.42 | 100 | 122 | 177 | 190 | 2 |
| Raby Castle | 3 | 3 | — | — | — | 100 | 118 | 177 | — | 1 |
| Red Lake | 2 | 2 | 2.8 | 13.0 | 0.52 | 93 | — | 177 | 190 | 3 |
| Rolan | 2 | 2 | 2.7 | 11.5 | 0.99 | 100 | 123 | 182 | 208 | 2 |
| Rondom | 3 | 2 | 2.7 | 11.0 | 0.48 | 104 | — | — | — | — |
| Rosetta | 2 | 2 | 2.8 | 9.5 | 0.92 | 96 | 118 | 187 | 200 | 2 |
| Rovada | 3 | 1 | 2.7 | 11.0 | 0.46 | — | — | — | — | — |
| Stanza | 2 | 2 | 2.7 | 10.5 | 0.84 | 94 | — | — | — | — |
| Stephens #9 | 2 | 3 | 2.9 | 11.5 | 0.99 | 96 | 125 | 183 | 200 | 2 |
| Viking | 2 | 1 | 3.0 | 10.0 | 0.47 | 100 | 120 | 187 | 194 | 2 |
| Welcome | 2 | 1 | 2.7 | 12.0 | 0.49 | — | — | — | — | — |
| Wilder | 3 | 2 | 2.8 | 12.5 | 0.58 | 93 | — | — | — | — |

*Note:* Mildew (American powdery mildew) resistance: 1 = poor, 2 = moderate, 3 = good, 4 = no symptoms; vigor: 1 = poor, 2 = moderate, 3 = good; picking ease: 1 = difficult, 2 = average, 3 = easy; bloom and ripening dates: 1 = January 1, 365 = December 31; data missing or not available: —.

## Recommended Red Currants

### Cascade

The fruits of 'Cascade' are large, high in sugar, and have good flavor. The canes are moderately vigorous and moderately resistant to powdery mildew. It was released by the University of Minnesota in 1942 as a seedling of 'Diploma'.

### Cherry

The fruit of the cultivar 'Cherry' ripens during the early season, varies from small to large, and has excellent flavor and good quality. Cluster stems are short and close to the wood, making hand harvesting difficult. The canes are vigorous and productive but have a tendency to form "blind wood" (does not form a terminal shoot). It was once the most popular red currant in America for the home garden and market. Although the fruit flavor and color are excellent, the difficulty in picking, nonuniform fruit size, and tendency to set blind wood limit its commercial value; however, it is an excellent choice for home use. Probably originating in Italy, it was introduced to France in 1840 and renamed 'Cherry'. Imported to Flushing, New York, by Dr. Wm Falk in 1846, Shoemaker (1948) reported that the 'Cherry' cultivar originally imported into North America had been confused with other cultivars and that the original cultivar had probably been lost.

### Diploma

'Diploma' berries are very large, juicy, and ripen over a short period during the mid-season. They are borne in loose, medium-length clusters that fill poorly at the tips. The canes are medium to large, vigorous, erect to spreading, productive, and resistant to powdery mildew. According to Hedrick (1925), it is one of the best red currants for home and commercial production. One problem with 'Diploma' is that the cluster stems are short and the fruit very tender, which requires that the berries be picked and handled with great care. It was developed in New York in 1885 as a cross between 'Cherry' and 'White Grape'.

*Fay*

Also known as 'Fay's Prolific', the fruit is large, firm, and juicy with a thin, tough skin and excellent quality. The berries ripen in the early mid-season. The bushes are compact and somewhat lacking in both vigor and productivity, which limits its commercial value. It was developed in New York in 1868, possibly as a cross between 'Victoria' and 'Cherry'.

*Jonkheer van Tets*

The berries of 'Jonkheer van Tets' are variable in size, of average quality, and have a tendency to split in wet weather. The canes are erect, vigorous, susceptible to wind damage, and only moderately resistant to mildew. In University of Idaho trials, 'Jonkheer van Tets' demonstrated only mediocre performance. It is included in this list because it seems to perform better in other growing regions of North America and has proven very popular in Europe. Likely to prove valuable as a parent in breeding red currants, it was developed by J. Maarse in the Netherlands and released in 1941 as a seedling of 'Fays Prolific'.

*Laxton's No. 1*

'Laxton's No. 1' berries are medium sized with small seeds and good overall quality. The fruit is tightly bunched on moderately long strigs. The canes are erect to slightly spreading and very productive. A popular commercial cultivar in Europe, it was developed by T. Laxton in England between 1890 and 1900.

*Minnesota No. 71*

The berries of 'Minnesota No. 71' are medium to large, have good, consistent quality, and ripen during the mid-season. The canes are vigorous, erect, and very resistant to powdery mildew. This is a popular cultivar for commercial production in England and is one of the best red currants grown in North America. It was developed by W. Alderman in Minnesota and released in 1933.

## Perfection

'Perfection' berries ripen during the early mid-season and are large, uniform, juicy, flavorful, and have a thin, tough skin. The cluster stems are long and easy to pick. The canes are small, vigorous, only moderately dense, and moderately resistant to powdery mildew. One of the most cold-hardy currants, it was developed in New York in 1887 as a cross between 'Fay' and 'White Grape' and released in 1902. Although the flavor, fruit quality, and productivity are excellent, the canes are susceptible to breaking and the fruit sunscalds easily if not picked promptly after ripening; however, it is possibly a good selection for commercial production in cooler northern areas.

## Portal Ruby

The berries of 'Portal Ruby' are very large and tart. The canes are moderately vigorous and moderately resistant to powdery mildew. It is a selection of *R. rubrum,* originally from England.

## Prince Albert

'Prince Albert' is sometimes confused with 'Red Dutch', but the two cultivars are different. The fruit is small to medium in size, ripens very late in the season, and hangs on the clusters for an unusually long time. The clusters are long, loose, well filled, and have long, easy-to-pick stems. The canes are moderately vigorous and quite resistant to powdery mildew. It has excellent potential for a late-season cultivar, particularly for U-pick operations. Introduced into the United States from France in 1850, its country of origin is unknown.

## Red Lake

'Red Lake' berries are medium to large, uniform, juicy, flavorful, and ripen during the mid to late season. The canes are moderately vigorous, erect, and moderately resistant to powdery mildew. The clusters are long and easy to pick. Developed by W. Alderman and introduced by the Minnesota Fruit Breeding Farm in 1933, it was tested as Minnesota 24 but its pedigree is unknown. It is an excellent choice for commercial and home production.

## Rondom

The fruit of 'Rondom' is small to medium sized, thick skinned, and ripens late in the season. The berries are clustered tightly on the strigs and hang well after ripening. The canes are erect, vigorous, and quite resistant to powdery mildew. Although this is a popular commercial cultivar in Europe, there is clonal variation within the cultivar and it is important to get plants that show the desired characteristics. Developed by J. Rietsema in the Netherlands and released in 1946, it is a backcross of *R. multiflorum* to red currant cultivars.

## Rosetta

'Rosetta' berries are large, tart, and ripen late in the season. The long, loose fruit clusters facilitate handpicking. The canes are moderately vigorous and moderately resistant to powdery mildew. A promising introduction from Holland, it does not have much of a track record yet in North America. It was released in the Netherlands by the Wageningen Agricultural University in 1974 from the cross 'Jonkheer van Tets' × 'Heinemann's Rote Spatlese'.

## Rovada

The dark red berries of 'Rovada' are large, have average flavor, and ripen late in the season, just before 'Rotet'. The fruit clusters are long and loose, facilitating handpicking, and are moderately susceptible to runoff. Resistant to leafspot, it is a popular variety for commercial production in Central Europe due to its excellent fruit quality and yields, but it is also suited for home gardens. Developed in 1980 in the Netherlands by L. M. Wassenaar of the Institute of Horticulture and Plant Breeding, it is the product of 'Fay's Prolific' × 'Heinemann's Rote Spatlese'.

## Stanza

The medium to large berries of 'Stanza' ripen during the mid to late season. They are borne on short strigs that have long, easy-to-pick stems. The canes are moderately vigorous and reliably productive. The fruit is suitable for fresh use and especially for juicing. The flavor is acidic. It was released in 1967 by the Fruit Growing Experiment Station in the Netherlands as a selection of *R. rubrum*.

## Stephens No. 9

The fruit of 'Stephens No. 9' is large to very large. The canes are vigorous and quite resistant to powdery mildew. It is one of the most popular red currants in North America. Bred by C. L. Stephens of Orilla, Ontario, in about 1933, its parentage is unknown.

## Tatran

'Tatran' canes are tall, vigorous, and noted for excellent coldhardiness. The berries are large, have good quality, are borne on exceptionally long clusters, and ripen late in the season. Yields are, reportedly, high. The cultivar is new to North America, but reports from the northeastern United States are favorable. Developed by J. Cvopa and I. Hricovsky at the Research Institute of Bojnice, Czechoslovakia, it was released in 1985 from the cross 'Red Lake' × 'Goppert'.

## Victoria

Reportedly synonymous with 'Wilson's Long Bunch', the berries of 'Victoria' are small to medium in size, firm, juicy, and have a tough skin. They ripen during the late mid-season, are borne on short to medium, loose clusters, and hang on the clusters for an unusually long time. The canes are large, vigorous, erect, dense, productive to very productive, and very cold hardy. 'Victoria' was developed about 1800 in England and has been confused, at times, with 'Raby Castle' and 'Houghton Castle'. At one time, it was called 'May's Victoria'. According to Hedrick (1925) if it were not for the small fruit, 'Victoria' would be one of the best red currants. With mechanical harvesting and a processing market, it may still be a good commercial choice.

## Viking

'Viking' berries are medium sized. The canes are less vigorous than other cultivars and are only moderately resistant to powdery mildew. This cultivar's strong point is that it has proven resistant, although not immune, to white pine blister rust. Developed in Norway in 1945, it is a seedling of *R. petraeum* and *R. rubrum,* although the direction of the cross is unknown.

*Wilder*

The berries of 'Wilder' are variable, tending to be small to medium in size. They are tender, juicy, have good quality, are borne on long, easy-to-pick clusters, and hang on the clusters for an unusually long time after ripening. The canes are vigorous, erect to slightly spreading, and very resistant to powdery mildew. 'Wilder' is a seedling of 'Versailles' and was selected in about 1877 by E. Y. Teas in Indiana. During the early 1900s, 'Wilder' was the leading commercial currant grown in New York and remains a good choice today. It was named after Marshall P. Wilder, former president of the American Pomological Society (Shoemaker, 1948).

## WHITE CURRANT CULTIVARS

Relatively few white currant cultivars are available. Germany and the former Czechoslovakia are the leading producers of white currants, which are primarily used for baby food (Harmat et al., 1990). The leading cultivars in those countries are 'Werdavia', 'Zitavia', 'Meridian', and 'Victoria'. In Britain, 'White Versailles' is popular. Newer white currants from Holland are gradually replacing older cultivars for European production (Brennan, 1996). In Northwest trials, 'White Imperial', 'White Currant 1301', and 'White Dutch' perform quite well. Of those, 'White Imperial' has been the most consistent performer and most resistant to powdery mildew. Darrow (1919) recommended 'Victoria' for the Pacific Coast region of the United States. 'White Grape' is generally considered to have inferior fruit quality, despite being very cold hardy. Where blister rust is of concern, 'Gloire des Sablons' proved immune or highly resistant to blister rust tests in Oregon. Table 7.4 gives the result of the authors' evaluations of selected white currants.

TABLE 7.4. White currant cultivar characteristics.

| Cultivar | Mildew resistance | Vigor | pH | Fruit TSS (%) | Weight (grams) | Bloom first | Bloom end | Ripening first | Ripening end | Picking ease |
|---|---|---|---|---|---|---|---|---|---|---|
| Gloire des Sablons | 2 | 3 | 2.9 | 12.5 | 0.66 | 95 | — | — | — | 2 |
| Rosa Hollandiscme | 3 | 3 | 2.7 | 11.0 | 0.63 | 100 | 133 | 187 | 196 | 3 |
| White Currant 1301 | 3 | 2 | 2.9 | 12.5 | 0.64 | 100 | 122 | 183 | 208 | 2 |
| White Imperial | 3 | 3 | 2.8 | 14.0 | 0.52 | 92 | 120 | 177 | 191 | 2 |
| White Versailles | 2 | 2 | 2.9 | 12.5 | 0.60 | 100 | 122 | 185 | 202 | 3 |

*Note*: Mildew (American powdery mildew) resistance: 1 = poor, 2 = moderate, 3 = good, 4 = no symptoms; vigor: 1 = poor, 2 = moderate, 3 = good; picking ease: 1 = difficult, 2 = average, 3 = easy; bloom and ripening dates: 1 = January 1, 365 = December 31; data missing or not available: —.

## Recommended White Cultivars

### Blanka

Noted as being late ripening, very productive, and very resistant to frosts, 'Blanka' was bred by J. Cvopa and I. Hricovsky and released in Slovakia in 1977 from the cross 'Heinemann's Rote Spatlese' × 'Red Lake'.

### Primus

'Primus' berries are somewhat smaller than those of 'White Imperial' and have good flavor. Fruit color is whiter than is found in most other white currant cultivars. Yields are reported to be high. It was bed by J. Cvopa and I. Hricovsky and released in Slovakia in 1977 as the product of 'Heinemann's Rote Spatlese' × 'Red Lake'.

### White Currant 1301

The berries of 'White Currant 1301' are medium sized and are borne on somewhat vigorous canes that are moderately resistant to powdery mildew. It was developed from the Swedish University of Agricultural Science as a white sport of *R. rubrum*.

### White Dutch

The berries of 'White Dutch' are small to medium in size, juicy, sweet, and have excellent quality. They ripen very early and are darker than other white cultivars. The canes are medium sized, vigorous, erect to slightly spreading, dense, very productive, and quite resistant to powdery mildew. 'White Dutch' is one of the oldest named white currants and has often been confused with other cultivars. It was developed before 1729 in the Netherlands and probably introduced into America in the early 1800s. The small, uneven size of the berries is the main limitation of this cultivar.

### White Imperial

'White Imperial' fruit varies in size, ranging from medium to large, and is juicy and tender. Hedrick (1925) rated its quality as very good to best of all white currants. The canes are medium sized, vigorous, spreading, productive, and very resistant to powdery mildew. Developed by S. D. Willard in New York in about 1890, 'White Imperial' was highly recommended in North America in the 1890s. It is an excellent choice for commercial and home production in North America.

### White Versailles

The fruit of 'White Versailles' is large, juicy, has good quality, and is borne on long clusters. The canes are vigorous and erect. It has long been a leading cultivar in Europe. Developed by M. Bertin in France prior to 1883, its parentage is unknown.

## JOSTABERRY CULTIVARS

Jostaberries are hybrids between gooseberries and black currants. The bushes are very vigorous and highly resistant to powdery mildew in northern Idaho trials. They are also, reportedly, resistant to blister rust. The fruits are intermediate in size between gooseberries and currants, resemble gooseberries in shape, are dark red to black, and lack a strong black currant flavor. Some people complain that the fruits lack the good characteristics of either black currants or gooseberries, while others like jostaberries for processing. Perhaps the greatest challenge a jostaberry farmer might face is marketing a fruit that virtually no one in North America has heard of.

Three jostaberry cultivars are cultivated. At present, only 'Josta' is widely available in North America. Although 'Josta' is reported to be self-fertile, North American growers have reported problems with poor fruit set. The cultivars Jostine and Jogranda (the latter also known as 'Jostagranda' and 'Jostaki') have been imported into the United States and are now available through the USDA-ARS National Clonal Germplasm Repository. These two cultivars are at least partially self-sterile and should be planted together for good fruit set. Table 2.3 lists the fruit characteristics for 'Josta'.

# Chapter 8

# Planting and Establishing *Ribes*

## *ORDERING PLANT MATERIAL*

Before ordering *Ribes,* U.S. growers should ensure their states allow the plants to be imported and grown. State departments of agriculture can provide information regarding quarantines. Plants from foreign sources must go through testing in national quarantine. Canada has no national or provincial restrictions on growing currants and gooseberries, although import certification and phytosanitary certificates are required as with any other nursery stock (Dale, 1992).

Nurseries sell the plants as either container-grown or fall-dug bareroot stock. Because *Ribes* break dormancy and begin growing very early in the season, nurseries seldom dig them during the spring. Most commercial growers prefer bareroot plants because they are less expensive to purchase and ship than containerized plants. Hobby growers generally find containerized plants more convenient than bareroot stock.

Bareroot and containerized plants sold by nurseries are generally one or sometimes two years old. Although either is acceptable, some authorities have expressed a preference for vigorous one-year-old plants, believing they are easier to transplant (Shoemaker, 1948) and are less expensive to purchase (Dale and Schooley, 1999). In either case, the plants should have from one to several pencil-thick canes 12 to 18 inches (30 to 45 cm) long. The root system should be dense and fibrous. Be especially careful to examine the roots of containerized plants. *Ribes* produce vigorous roots and can quickly become root bound. Plants that have remained in containers too long will have many circling roots. Such plants dry out and wilt easily, and normally do not perform well when transplanted to the field. Currants, gooseberries, and jostaberries are not good candidates for long-term container culture.

Planting stock should be healthy and free from pests and diseases. Reject those whose leaves or canes show dead, deformed areas or masses of white or brownish-gray fungus (powdery mildew). One of the most important steps in avoiding pest and disease problems is to start with clean, healthy stock.

Because of *Ribes'* early spring growth, bareroot stock is usually fall planted. An exception to this rule is in areas with cold, dry winters, such as the Great Plains (Card, 1907). In such areas, growers may find it advantageous to plant in the spring.

Bareroot stock can be overwintered in cold storage, provided that the temperature is kept close to 32°F (0°C) and the stock is planted early in the spring. Planting bareroot stock in late spring can seriously interfere with that season's growth (Shoemaker, 1948). Containerized stock can be planted any time from spring through fall, although the earlier the bushes are planted, the better they will establish before winter. In areas subject to frost heaving or cold, dry winter winds, growers may find it beneficial to mulch summer- and fall-planted bushes with sawdust or bark chips. Straw mulches provide excellent habitat for mice and other rodents that can girdle canes.

## PLANTING

For centuries, selecting alley widths and in-row planting distances for currants and gooseberries was simple and straightforward. Although alley widths increased from about six feet (2 m) to as much as 10 to 12 feet (3 to 3.7 m) as farmers switched from horse-drawn implements to tractors, spacings of plants within rows remained relatively unchanged. At the closest spacings, a one-acre field might contain 3,600 plants (8,892 plants/ha), although more typical spacings required 1,300 to 1,400 plants per acre (3,211 to 3,458 plants/ha).

Following World War II, labor shortages and the cost of labor increased in Europe. With handpicking representing a large percentage of currant and gooseberry production costs, efforts began to develop mechanical harvesters. During the early 1970s, mechanical currant and gooseberry harvesters became commercially available in Europe. It soon became apparent that new planting and training systems were needed to accommodate the mechanical harvesters and take advantage of opportunities for increased yields. While typical black currant fields designed for mechanical harvest might contain about 1,740 plants/acre (4,300 plants/ha) (Harmat et al., 1990), one research trial on high-density black currant production used 16,000 plants/acre (40,000 plants/ha) (Olander, 1993). Growers also found that they needed to leave more room at the ends of rows to allow tractors and mechanical harvesters to turn around. Complicating the situation is the diversity of harvesters, ranging from handheld vibrating shakers, to half- and full-row tractor drawn machines, to large, self-powered, over-the-row harvesters.

The following recommendations for handharvested fields work well for most growing regions. Information on planting systems for mechanical harvesting is necessarily general. Growers planning to harvest their currant or gooseberry crops mechanically should decide on a particular machine and base planting distances and alley widths on the manufacturer's recommendations.

Two distances must be kept in mind when laying out a field. The first is the distance between rows, and the second is the distance between plants. Row spacing depends on the equipment used to tend the planting. With a small tractor in the 15 to 30 horsepower range, eight to nine feet (2.5 to 2.7 m) is about as close as one can plant and maneuver between mature bushes without damaging them. For larger tractors or when mechanical harvesters will be used, space the rows 10 to 12 feet (3 to 3.7 m) apart.

Red currants, white currants, and gooseberries are normally planted three to four feet (1.0 to 1.2 m) apart within rows (Harmat et al., 1990), although some authors (Dale and Schooley, 1999) recommend closer spacings of two to four feet (0.6 to 1.2 m). Compact gooseberry cultivars often perform well at the closer spacing, while vigorous red currant cultivars such as 'Minnesota 71' or 'Wilder' are best planted at the wider spacings. Jostaberries are the most vigorous cultivated *Ribes* and perform well when planted at least six feet (2 m) apart within rows. Crowding the bushes within rows can increase pests and diseases. Unless a close spacing is needed for mechanical harvesting, wider spacings provide better light and air penetration into the bushes, reducing pest and disease problems.

Black currants are more vigorous than red and white cultivars and are generally planted farther apart. For hand harvesting, space black currants four to five feet (1.2 to 1.5 m) apart (Harmat et al., 1990). When fields are to be mechanically harvested, set two-year-old plants 2.5 to 3 feet (0.8 to 1.0 m) or unrooted cuttings one foot (0.3 m) apart

(Harmat et al., 1990). Black currants can also be grown in hedgerows and cropped in an alternate-year system, as discussed in detail in Chapter 9. When creating a hedgerow, plant black currants two feet (0.6 m) apart within the rows. A grower might also choose to plant black currants four to five feet (1.2 to 1.5 m) apart within rows to save money on nursery stock. During the first dormant-season pruning, cuttings from one-year-old canes are cut into six-inch (15 cm) lengths and stuck base down into the soil between the bushes. Many of the cuttings will root and form a hedgerow within one to two years.

In high-density trials of black currants, Olander (1993) set plants 18 inches (46 cm) apart within and between rows. The method is similar to meadow orchard systems in apple and seeks to maximize fruiting wood. With the spacing tested, yields were doubled compared with conventional row systems. Harvest can begin as early as the year after planting but normally begins in the plantation's third growing season. Harvest consists of using a combine-style mechanical harvester to cut all canes off just above the ground. Inside the harvester, the berries are shaken from the canes and collected. In this alternate-year system, nonfruiting canes grow the first year and are harvested as they bear fruit their second year. Not all cultivars are suited to this high-density system. 'Ojebyn' and other selections from Balsgard proved sensitive to shading and produced low yields. 'Ben Alder' and 'Ben Sarek', on the other hand, bore fruits along the lengths of their second-year canes. With this production system, yields can be increased, at least for the short term, and hand-pruning and handpicking labor are eliminated. Cost of establishment is very high; weed, pest, and disease control can be difficult; and a specialized harvester is needed.

The number of plants needed for a field is calculated by dividing the total planting area by the space required for each bush. Say, for example, that the planting area is exactly one acre (43,560 square feet) and rows will be spaced ten feet apart with bushes set five feet apart within rows. The space required for each bush is $10 \times 5 = 50$ square feet. Dividing the planting area of 43,560 square feet by 50 square feet per bush shows that 871 plants will be needed. When designing a planting, allow for roads and equipment turnaround space at the ends of the rows.

When planting bareroot stock, carefully inspect the roots and use sharp pruners to remove any dead or diseased roots. Also prune off thick woody roots that are kinked, twisted around the collar, or point inward toward the collar. For container stock, cut through circling roots around the outsides of the rootballs using a sharp knife.

For small plantings using bareroot stock, dig planting holes about 12 inches (30 cm) deep and 18 inches (45 cm) in diameter. Build a cone of soil about eight inches (20 cm) high in the bottom of each hole and spread the roots evenly over the cones, letting them drape toward the bottoms of the holes. Set the bushes somewhat shallower than they grew in the nursery to allow for settling when filling the holes. Prune back any roots that do not fit into the holes. Fill the holes using only the soil that originally came out of them—do not add any soil amendments. Doing so can interfere with moisture moving into and out of the root zone. A far better practice is to apply organic amendments in a layer over a field and till the amendment into the soil before planting. After a plant has been set into the ground, firm soil over the roots by hand and settle the soil into place by soaking it several times with water. After settling, the bushes should rest at about the same depths as they grew in the nursery. For containerized plants, dig holes large enough to hold the root balls and plant as described. Water heavily immediately after planting.

For larger plantings, digging individual holes can be expensive and time-consuming. A more cost-effective method is to plow 12-inch-deep (30 cm) furrows centered on the plant rows and to set the plants into the furrows. Take care to spread the roots along the furrows and to firm the soil around each plant.

Some growers prefer to cut the canes on newly planted black currants so as to leave only two buds above ground on each cane. Shortening the canes can help stimulate the development of new shoots from below the ground, creating a plant with multiple stems, rather than a single trunk.

Applying sawdust, bark chips, or other nonstraw organic mulches around newly planted bushes helps control weeds and keeps the soil cool. Dale (2000) found that growing black currants through black plastic mulch and keeping alleys cultivated significantly increased yields over treatments that did not use cultivation and plastic mulch. With this system, *Ribes* can be grown on level ground or on hills about 18 inches (0.5 m) wide and 12 inches (30 cm) high. In either case, a drip irrigation tube should be run along the crop row before the plastic film is installed.

In strawberry "plasticulture" the strawberries are planted on hills and plastic film is stretched over the plants and held in place by burying the edges of the plastic under soil at the bottoms of the hills. Disks of film are then removed over each strawberry plant by burning through the plastic with a hot metal cylinder attached to the end of a wand fitted with a propane-fired nozzle. The same approach can be used for currants and gooseberries, although it may be necessary to shorten canes to three to six inches (7 to 15 cm) before applying the plastic. Using a burner is faster and less likely to tear the plastic film than cutting holes with a knife.

Although currants and gooseberries are normally not fertilized at the time of planting, extra nitrogen may be needed if uncomposted organic mulches are applied around the plants. A rule of thumb would be to apply one-half ounce of actual nitrogen per cubic foot (500 g/m$^3$) of mulch. Directions for determining how much of a particular fertilizer to apply are given in Chapter 6.

# Chapter 9

# Crop Management

## *NUTRITION AND FERTILIZATION*

Plants, like all living things, take inanimate materials from the surrounding environment and use them to create living tissues. These inanimate materials are usually referred to as "essential elements." They are called essential because plants must have them in order to complete their life cycles. At present, 16 elements are considered necessary for normal plant growth and development. Three of these, carbon, hydrogen, and oxygen, are obtained from water and air, and make up the bulk of plant tissues in the form of carbohydrates. Typical carbohydrates include cellulose, starches, and sugars.

The remaining 13 essential elements are derived from soil and are normally referred to as mineral nutrients. They are divided into two categories, macronutrients and micronutrients. Plants need relatively large amounts of macronutrients, which are normally measured as percentages, by weight, in plant tissues. Micronutrients are as important to plant growth and development as macronutrients but are needed in minute quantities, generally on the order of parts of nutrients per million parts of plant tissue. Table 9.1 lists plant nutrients and typical concentrations in currant and gooseberry tissues.

Nitrogen is the nutrient most often deficient in farm and garden soils, partly because plants need large amounts of nitrogen and partly because nitrogen normally does not bind to soil particles and washes out of the root zone. Nitrogen is contained in protein molecules that combine to form all enzymes and many structural components of plant cells. Plants deficient in nitrogen are generally weak, spindly, yellowish (chlorotic), and unproductive. Although some plants, such as blueberries, respond best to a particular form of nitrogen, *Ribes* adapt well to various forms of the nutrient.

TABLE 9.1. Recommended nutrient levels in leaves of currants, gooseberries, and jostaberries.

| Nutrient | Deficient | Adequate | Excessive |
|---|---|---|---|
| *Macronutrients (% dry weight)* | | | |
| Nitrogen (N) | <2.6 | 2.7-2.9 | >3.0 |
| Phosphorus (P) | <0.25 | 0.26-0.30 | >0.30 |
| Potassium (K) | <1.0 | 1.0-1.6 | >1.6 |
| Calcium (Ca) | — | 1.0-1.5 | — |
| Magnesium (Mg) | <0.10 | 0.10-0.15 | >0.15 |
| *Micronutrients (ppm dry weight)* | | | |
| Manganese (Mn) | — | 20-70 | — |
| Iron (Fe)[a] | — | — | — |
| Copper (Cu) | 5 | 5-20 | — |
| Boron (B) | <20 | 20-40 | >40 |
| Zinc (Zn) | <15 | 20-50 | — |

*Note:* These are approximate standards for currant, gooseberry, and jostaberry leaf tissue nutrient concentrations, based upon work done with *Ribes* and other fruit crops (Bould, 1969; Bradfield, 1969; Vang-Peterson, 1973; Westwood, 1978; Dow, 1980; Pritts et al., 1992).
[a]Foliar concentrations are often poor indicators of iron status in fruit plants (Barney et al., 1984). Visual deficiency symptoms are often better indicators.

Nitrogen differs from all other mineral nutrients in that it does not originate in the soil but in the atmosphere. Unlike carbon, however, which plants absorb directly from the air as carbon dioxide, nitrogen must be "fixed" into a form that plants can absorb through the roots. Leguminous plants (beans and peas, for example) and certain other species form symbiotic relationships with soil microorganisms and between the two are able to utilize atmospheric nitrogen directly. *Ribes* do not form such symbiotic partnerships and require that nitrogen be added to the soil. Nitrogen is an essential element, but too much of it can force excessive vegetative growth that interferes with fruit formation and increases the plant's susceptibility to pests and diseases.

Next to nitrogen, potassium, also known as potash, is the nutrient most often deficient in *Ribes* soils. According to Keeble and Rawes

(1948, p. 240), an "adequate supply of potash in the soil is of first importance" for gooseberry production. *Ribes* crops also need phosphorus in relatively large amounts. Other essential elements are generally available in sufficient amounts from the soil without fertilization but can be limiting on certain sites. Such deficiencies are normally detected during preplant soil tests and are corrected during field preparation. Nutrition must be balanced because nutrients in plants influence one another. Adding too much calcium or phosphorus, for example, can trigger the onset of iron chlorosis, even though iron is plentiful in the soil. Table 9.2 lists essential nutrients, along with the plant symptoms that indicate imbalances and steps one can take to correct the imbalances.

It is a common belief that currants and gooseberries do not need fertilization, probably because neglected or abandoned bushes often survive for many years. On the contrary, they must be fertilized regularly to ensure good vigor and production. Fall-planted bushes do not need fertilization at the time of planting. Provided that a site has been properly prepared, fertilizer may not be needed the following spring. Commercial growers normally begin fertilizing the first or second spring after fall planting. Spring- or summer-planted stock may benefit from fertilization soon after planting. Soil analyses and observing plant vigor provide clues in deciding when to begin fertilizing.

Traditionally, manure was the material of choice for adding nitrogen to the soil and is still valuable for that purpose. Manure has advantages over industrial fertilizers in that it also adds organic matter to the soil and the nitrogen is released slowly over a period of months or years. Manure has drawbacks, however. On modern farms where livestock and crops are seldom raised together, obtaining enough manure to fertilize crops can be difficult and expensive. Manure should be composted before applying it to the soil in order to kill weed seeds and wash out excessive salts.

Many farmers today use ammonium sulfate, urea, or "complete" fertilizers (those containing nitrogen, phosphorus, and potassium) to add nutrients to their fields. If the soil and/or irrigation water are alkaline, consider using ammonium sulfate or other acidifying fertilizers to manage soil pH. Several methods are available to determine what kind of and how much fertilizer to apply.

TABLE 9.2. Nutrient deficiency and toxicity symptoms and treatments.

| Element | Symptoms | Recommendations |
|---------|----------|-----------------|
| *Macronutrients* | | |
| Nitrogen | Small, yellowish-green foliage. Poor vigor and stunted growth. | Apply nitrogen in two or three split applications during May through July. See rate recommendations in Table 9.3. |
| | Very dark, green leaves. Excessive lush, succulent growth. Poor fruit set and large, soft, poorly colored berries. Excessive late growth and failure to acclimate in the fall. | Excessive amounts of N are being applied. Reduce N fertilizations. |
| Phosphorus | Purple coloration appearing on older foliage. Premature leaf abscission. | 300 lb/acre (340 kg/ha) superphosphate, 150 lb/acre (170 kg/ha) triple phosphate, or 250 lb/acre (280 kg/ha) bone meal. Apply in early spring or late fall. |
| Potassium | Scorching of leaf margins and necrotic spots on older leaves. Bronzing of leaves. Premature coloring of undersized fruit. Weak growth and premature defoliation. | 400 lb/acre (450 kg/ha) sul-po-mag or 160 lb/acre (180 kg/ha) potassium sulfate. Apply when symptoms develop. Potassium should normally be applied during early spring or late fall. |
| Magnesium | Yellowing between veins of older leaves, with necrotic spots developing later. Premature leaf abscission. | 500 lb/acre (560 kg/ha) magnesium sulfate (epsom salts) or potassium magnesium sulfate (sul-po-mag) when symptoms develop or in early spring. Apply to soil, not to foliage. |

| Calcium | Leaf tips become scorched. Blossom ends of berries may rot. | Add 1,000 to 2,000 lb/acre (1,125-2,250 kg/ha) of actual calcium in spring or whenever symptoms develop. Calcium sulfate (gypsum) gives more rapid results than lime. |

*Micronutrients*

| Boron | Delayed bud break and elongated, slender leaves. Shoots and buds die back. | 1.5 lb/acre (1.7 kg/ha) solubor or borospray as a foliar spray any time leaves are present. Applications in early spring are best. A spring application to the soil of 175 lb/acre (197 kg/ha) borated gypsum or 10 to 15 lb/acre (11 to 17 kg/ha) of borax should prevent any boron deficiencies. *Caution:* Boron becomes toxic to plants at very low concentrations. Never band dry boron fertilizers within crop rows. Use soil and foliar tests to determine the need for boron. Do not apply more than one to two lb/acre of actual boron. Follow label directions when applying foliar spray materials. |

| Iron | Young terminal leaves turn yellow or white between green veins. Leaf margins and interveinal areas become brown and necrotic. Shoot growth is stunted. Fruit set is greatly reduced. Symptoms typically appear in spring or when soils are cold and wet. Most common on alkaline sites with soil pH values greater than 7.0. | Apply 2 lb/acre (2.3 kg/ha) iron chelate as a foliar spray when symptoms appear. Follow label directions. *Note:* Iron chlorosis is often caused by overirrigation or poor drainage, especially on alkaline soils. Ensure that irrigation water is evenly distributed throughout field. Drain or do not plant on wet sites. Use sulfur to lower soil pH to 6.5 on alkaline sites. |

TABLE 9.2 (continued)

| Element | Symptoms | Recommendations |
|---|---|---|
| Manganese | Dull interveinal chlorosis in older leaves. Young terminal leaves remain green. | Foliar sprays of 2 lb/acre (2.3 kg/ha) manganese sulfate or use manganese chelate per label directions as soon as leaves are well developed. High soil pH can create manganese deficiency. Use sulfur to lower soil pH to 6.5 on alkaline sites. |
| Zinc | Leaves are small and narrow with a striped, irregular chlorotic pattern between green veins. Rosettes of leaves form at the tips of shoots, with bare wood below the rosettes. Foliar analysis is not necessarily a good indicator of zinc status in plants. | Apply 44 lb/acre (50 kg/ha) of zinc sulfate in a spray to dormant bushes or 22 lb/acre (25 kg/ha) as a foliar spray after harvest. Commercial formulations of chelated zinc can also be used as foliar sprays. Follow label directions. Zinc can also be applied to the soil. Use 10 to 30 lb/acre (11 to 33 kg/ha) of zinc sulfate, or use chelated zinc according to label directions. |

Note: To convert from lb/acre or kg/ha to ounces or grams of fertilizer per plant, divide the figures in Table 9.2 by the numbers of plants per acre or hectare given in Table 5.1 and 5.2 for various row and in-row plant spacings. Then multiply lb/plant by 16 to calculate ounces per plant, or multiply kg/plant by 100 to calculate grams per plant.

## Methods to Determine Fertilization Needs

### Rule of Thumb

With this method, one simply adds the amount of fertilizers that other farmers have found to be adequate. Although the method is not absolutely precise, it generally works well, at least for a starting point. Table 9.3 provides rule of thumb recommendations. By following these recommendations, *Ribes* crops will probably receive adequate fertilization, although it is possible to apply excessive amounts or unnecessarily expensive fertilizer blends than are needed. Commercial growers, especially, are advised to fine-tune fertilizer selection and application using soil and foliar analyses.

### Visual Symptoms of Deficiencies

Although it is important to monitor crops regularly for problems, including signs of nutrient deficiencies, this should not be the pri-

TABLE 9.3. Recommended amounts of commonly available fertilizers to apply annually to *Ribes* crops. This table serves as an example only. Other manures and commercial fertilizer formulations can also be used to apply needed nutrients.

| Year | Composted manures pounds (kg) per bush | | Commercial fertilizers ounces (g) per bush | | | |
|---|---|---|---|---|---|---|
| | cow or horse | rabbit or poultry | 10-10-10 | 18-5-10 | 21-0-0 | 46-0-0 |
| 1 (planting) | 5 (2.5) | 1.5 (0.6) | 4 (115) | 2.2 (64) | 1.9 (55) | 0.9 (25) |
| 2 | 5 (2.5) | 1.5 (0.6) | 4 (115) | 2.2 (64) | 1.9 (55) | 0.9 (25) |
| 3 | 8 (3.5) | 2.0 (1.0) | 6 (170) | 3.3 (94) | 2.9 (81) | 1.3 (37) |
| 4 | 10 (4.5) | 3.0 (1.3) | 8 (225) | 4.4 (125) | 3.8 (107) | 1.7 (49) |
| 5+ | 13 (6.0) | 3.5 (1.7) | 10 (285) | 5.6 (158) | 4.8 (136) | 2.2 (62) |

*Note:* 10-10-10 contains 10 percent each nitrogen (N), phosphorus ($P_2O_5$), and potassium ($K_2O$). 18-5-10 is often formulated as a slow-release fertilizer containing 18 percent N, 5 percent $P_2O_2$, and 10 percent $K_2O$. Ammonium sulfate (21-0-0) and urea (46-0-0) contain 21 percent and 46 percent N, respectfully, but no P or K. Cow and horse manures contain approximately 0.5 percent N. Poultry and rabbit manures contain approximately 1.8 percent N.

mary method of determining fertilization needs. By the time a plant shows visible nutrient deficiencies, yields and plant health have already been adversely affected.

## Soil Analyses

Chemical analyses of cropland soil are important, especially before planting. The analyses provide a starting point for amending the soil. Soil analyses on a regular basis are also important for monitoring pH and organic matter content, both of which can affect nutrient uptake by plants. For mature woody perennial plants, however, soil analyses do not give accurate indications of the nutritional status of the plants themselves. Woody plants accumulate, store, and recycle nutrients within their tissues at higher concentrations than are present in the soil. After planting, soil analyses are normally used to monitor pH and organic matter every two or three years.

## Tissue Analysis

Chemical analyses of leaves and other tissues give the most accurate picture of the nutritional status of a plant because one is looking at what the plant actually contains. Commercial fruit growers commonly use tissue analyses to monitor the nutritional status of their crops. In this way, they can detect trends and take action early enough to avoid deficiencies and subsequent reductions in plant health, yield, and fruit quality. Unfortunately, nutrition in *Ribes* has been studied far less than for other fruit crops and specific recommendations are limited. Table 9.1 lists approximate standards for *Ribes* leaf tissue nutrient concentrations, based on work done with *Ribes* and other fruit crops. Because nutrient levels fluctuate throughout the growing season, the recommendations in this table assume that leaves are collected approximately the first of August. Always collect a combined sample of leaves from at least several bushes throughout the plantation. Within reason, the more bushes sampled, the greater the likelihood of determining an accurate estimation of a crop's nutritional status. In practice, collecting leaves from thirty individual bushes uniformly scattered throughout a field provides an adequate sample. Large fields should be divided into sections, with independent samples taken from each section. Analytical laboratories generally provide their clients with details on how to collect and prepare samples.

The amount of fertilizer needed depends on the age of a *Ribes* crop and the availability of nutrients in the soil. Young bushes need smaller amounts of nutrients than large, mature plants. As a general rule, mature *Ribes* plantings traditionally receive approximately 90 pounds of nitrogen, 18 pounds of $P_2O_5$, and 36 pounds of $K_2O$ per acre per year (100 kg N, 20 kg $P_2O_5$, and 40 kg $K_2O$ per ha). These rates work out to approximately four to eight ounces (110 to 220 g) of 10-10-10 fertilizer applied to each mature bush each spring (Harmat et al., 1990).

These fertilizer rates usually give good results, but they should be considered only as starting points. Aaltonen and Dalman (1993) examined single and split fertilizer applications in red and black currants for eight years. Red currant yields were unaffected by rates ranging from no fertilizer to 61 pounds per acre (0 to 71 kg/ha) of nitrogen. Similarly, nitrogen fertilizer rates from 22 to 40 pounds per acre (25 to 45 kg/ha) produced foliar and soil nutrient concentrations within the range recommended in Finland. Keeping accurate records of fertilizer applications, fruit yields, and tissue analyses helps growers refine fertilizer usage for their particular sites. Excessive fertilization is unnecessarily expensive and can create environmental problems. Fertilizing too little, however, can eventually "mine" nutrients from the soil and lead to poor yields and plant health.

Fertilizers containing about 10 percent each of nitrogen, $P_2O_5$, and $K_2O$ (10-10-10) are commonly recommended for *Ribes* crops in North America, although some growers prefer a ratio nearer 15:5:10. If soil and foliar phosphorus and potassium concentrations are adequate, nitrogen-only fertilizers are preferable. Table 9.3 lists recommended amounts of commonly available fertilizers that can be used over the life of a planting. Remember to add extra nitrogen if uncomposted organic mulches are used, as described at the end of Chapter 8. For growers who wish to apply fertilizers through irrigation lines (fertigation), soluble fertilizers are available. Commercial growers should select the most cost-effective fertilizers available to them, taking crop nutritional needs, material costs, and application and other labor costs into consideration. Certified organic growers will need to select fertilizers approved by their respective regulatory agencies.

Splitting fertilizer applications can reduce the amounts of nutrients leaching from the root zone. One common method is to apply one-third of a fertilizer in the early spring just as the new growth is forming, one-third during active growth, and the final third when the fruit begins ripening. Manures and commercial slow-release fertilizers release nutrients into the soil slowly over time and are best applied all at once, rather than in split applications.

## WEED CONTROL

Eliminating weeds prior to planting and keeping *Ribes* fields free of weeds are absolutely necessary to obtain good establishment and maintain healthy, productive plants. *Ribes* crops are shallow rooted and do not compete well with weeds for water and nutrients. Weeds also interfere with disease control, pest management, and harvesting. At a minimum, maintain three-foot-wide (1 m) weed-free strips centered along the berry rows.

Different weeds respond differently to weed-control programs. Some weeds, for example can be controlled with mowing, while others thrive under the same treatment. Likewise, some weeds can be easily controlled with one herbicide, although a different herbicide may be needed for another kind of weed. To develop an effective weed-control program, growers need to identify problem weeds that are present in their areas and learn to recognize and control them. In general, programs that utilize several different control methods are more effective than those based upon a single practice.

### Cultivation

Cultivation can effectively control some weeds, and recent research strongly supports the use of cultivated alleys in black currant fields (Dale, 2000). Because *Ribes* are shallow rooted, cultivation close to the bushes must also be shallow, no more than about one to two inches (2.5 to 5 cm) deep. Some earlier writers suggested that the roots may "be forced to go somewhat deeper by practicing fairly deep cultivation from the start" (Sears, 1925, p. 242), but no evidence supports this practice and we do not recommend it. The advantages of cultivation are that it does not add chemicals to the environment and effectively controls annual weeds. The disadvantages, if it is used excessively or otherwise improperly, are that cultivation can damage crop roots, break down soil structure, increase erosion, create dust that contaminates the berries, and increase soil compaction.

Because *Ribes* bushes are planted close together within rows, mechanical cultivation is normally limited to the alleys between rows. When horses and hand-powered tillers were used, the plants were often set in a cross pattern and at spacings that allowed cultivation both along rows and between bushes (Sears, 1925). With the tractors and cultivating implements used today, this practice is less popular than it once was. An exception might be the use of a walk-behind rototiller or similar implement for cross cultivating between young bushes. As the bushes mature and fill in the canopy, mechanical cultivation will be restricted to alleys. Tractor-mounted rototillers, grape hoes, discs, rakes, harrows, and similar devices are often used to control alley weeds in *Ribes* fields. Mechanical cultivation is generally supplemented with mulches, hand weeding, and/or herbicides within rows. Cultivation can also be used as part of systems that utilize annual cover crops in alleys and mulches and/or herbicides to manage weeds within the berry rows. In Europe, herbicides that keep alleys free of vegetation have largely replaced mechanical cultivation (Harmat et al., 1990).

## Mulches

### Organic Mulches

As mentioned previously, mulching crop rows with organic materials after planting can help manage weeds. Mulches generally act by excluding light from the soil surface. The seeds of many weed species must be exposed to light in order to germinate. Mulches also help to keep soil temperatures cool, conserve soil moisture, and add nutrients and organic matter to the soil, all of which are important in *Ribes* production (Van Meter, 1928; Harmat et al., 1990). Some organic mulches also provide favorable environments for beneficial insects and microorganisms.

Although organic mulches can be valuable additions to a *Ribes* operation, they have disadvantages as well. Mulches do not provide complete weed control, and organic materials contaminated with weed seeds, insects, and diseases can actually increase problems. Perennial weeds, such as quackgrass and Canada thistle, are especially troublesome in mulched fields because of their rhizomes. Rhizomes can spread quickly in some organic mulches, particularly sawdust or compost. *Ribes* roots can grow up into mulches, making hand cultivation difficult. If applied excessively deep, organic mulches can also harbor mice and voles.

The shallow root systems that form under organic mulches may also be susceptible to freezing injury during the winter and can make the bushes susceptible to drought stress. Sawdust, in particular, can create problems when used as a mulch. When dry, it repels water and interferes with irrigation. Sawdust can also provide a suitable environment for the germination and establishment of some weed seeds.

Bark and wood chips help control weeds with fewer problems than fine-textured sawdust, although research in Sweden showed that, over a four-year period, a bark chip mulch did not add organic matter or biomass to the soil and caused nitrogen deficiency in black currants, despite applications of 178 pounds of nitrogen per acre (400 kg/ha) in manure (Larsson et al., 1997).

Straw mulches also help control annual weeds, reduce soil temperatures, and maintain soil moisture, but they provide ideal habitats for rodents that can girdle canes. Use of straw mulching makes a rodent-control program advisable. Although a good source of plant nutrients, compost provides excellent conditions for weed seed germination and growth and may not give effective weed control when used for mulch.

Organic mulches for commercial plantings can be prohibitively expensive to purchase, transport, and apply. Applying four inches (10 cm) of sawdust to three-foot-wide strips with rows on nine-foot centers requires 180 cubic yards (137 m³) of sawdust weighing 29 tons (26 tonnes) per acre.

Excessively deep mulches encourage collar rot and inhibit the movement of oxygen into the soil. Apply mulches no more than about four inches (10 cm) deep and keep the mulches pulled away from the canes and collars of the plants to reduce disease problems.

Growers intending to apply organic mulches to *Ribes* fields should take great care to eliminate perennial weeds (particularly rhizomatous weeds) before planting their crops. Translocatable herbicides, such as glyphosate, often provide the easiest and most thorough weed control in perennial crops because they move from the shoots and leaves into the roots, killing the entire plant. Some contact herbicides kill only shoots and leaves, which resprout from undamaged roots and rhizomes. Organic growers utilize rotation crops, hand weeding, and cultivation to eliminate and manage perennial weeds.

## Plastic and Fabric Mulches

When *Ribes* are grown as individual bushes or cordons, plastic films or weed barrier fabrics can be used within crop rows to control weeds. Thousands of acres of strawberries are grown through plastic film in the United States. Growers typically install plastic drip irrigation lines under the films to apply irrigation water and liquid-soluble fertilizers, strategies that also work for currants, gooseberries, and jostaberries.

Black plastic films are more effective in controlling weeds and keep the soil cooler than clear films. In Canada, research with black currants showed that black plastic mulch within the berry rows increased yields by 26 percent over treatments with no plastic mulch. A combination of cultivation in the alleys and black plastic within the rows increased yields by 68 percent over a red fescue cover crop in the alleys and no plastic mulch within the berry rows (Dale, 2000).

Weed barrier fabrics provide partial weed control, although aggressive weeds can penetrate some fabrics from below. Many fabrics are also covered with organic mulch to protect the fabrics from ultraviolet light and mechanical damage. Weed seeds are able to germinate on the fabric under the mulch, with the roots sometimes penetrating through the fabric into the soil.

## Cover Crops

Cover crops planted in the alleys between bushes have been used in *Ribes* fields for more than a century. Cover crops are planted only in the alleys, leaving a three- to four-foot-wide (1 to 1.3 m) strip of weed-free soil centered along the berry rows. Cover crops help manage weeds, reduce soil erosion and compaction, improve movement of water and oxygen into the soil, and form a good working surface. Other advantages cover crops have include reducing root damage from cultivation and reducing dust contamination of the fruit. Some cover crops also provide a favorable environment for predatory insects and mites. Leguminous cover crops, such as clover and peas, fix atmospheric nitrogen that eventually becomes available to the crop plants. The use of cover crops helps to maintain high levels of soil organic matter, as opposed to clean cultivation, which decreases organic matter in the soil.

Negative aspects of cover crops are that they compete with crops for nutrients and water, require maintenance, and can invade crop rows. Dale (2000) found that black currant yields were reduced when red fescue was grown as an alley cover crop, as compared with cultivated alleys. Cover crops can also harbor insects, rodents, slugs, and diseases that damage fruit crops. Cover crops tend to increase the humidity below and within the canopy of the bushes, increasing the likelihood of fungal disease problems. Alfalfa and clovers may serve as reservoirs of alfalfa mosaic virus, which causes interveinal white mosaic in *Ribes*. Other broadleaf plants can serve as hosts for viruses listed in Table 10.1.

Cover crops can be either permanent or may be grown and subsequently incorporated into the soil on an annual basis. Permanent cover crops offer the advantages of always providing good working surfaces and eliminating the need to cultivate alleys. They can be planted at the same time as the *Ribes* or may be established later. If the alleys are not irrigated, cover crops should be planted in late fall or early spring to take advantage of seasonal rains. Most permanent cover crops must be mowed several times during the growing season and may need to be irrigated and fertilized. Permanent cover crops might be desirable in a U-pick operation, provided the grower considers customer convenience and comfort more important than maximum fruit yields.

The characteristics to look for in a permanent cover crop are that it be very durable and resistant to wear from machinery and foot traffic. It must be noninvasive, or it can move into crop rows, making hand weeding and/or herbicide application necessary. It should be low growing, require little mowing, and have low requirements for water and nutrients. Ideally, it should go dormant during the midsummer when the *Ribes* bushes are actively growing.

Certain fescues make good permanent cover crops, particularly hard and sheep fescues. The major drawbacks to hard and sheep fescues are that they are slow to establish and are bunch forming, rather than sod forming. They are normally sown with a nurse crop of more rapidly establishing grass, such as annual rye. On soils prone to frost heaving, sod-forming grasses can be more durable than bunch-type grasses, but are also more likely to invade the crop rows. Tall fescue is vigorous and deep rooted, requiring relatively frequent mowing and creating greater competition with crops. Dwarf white and strawberry clovers can make good permanent cover crops, especially when planted together with a grass. Note, however, that some clovers and other legumes have fleshy roots that are attractive to gophers. Some clovers can invade crop rows.

Annual cover crops are normally planted in the summer or fall, after the fruit is harvested. The cover crop is allowed to remain in place until spring and is then incorporated into the soil by shallow cultivation. Annual cover crops do not provide a permanent working surface, but they also require less care than permanent sods. Incorporating annual cover crops as green manures helps to maintain soil organic matter.

The characteristics to look for in an annual cover crop are that it establish rapidly, require little care, and not be invasive or form a dense sod. Summer annuals that are killed by winter cold make especially good cover crops. Spring barley, planted in the fall for a cover crop, has been used effectively in berry plantings. In the spring, the dead barley stubble can be mowed and left as a mulch, rather than incorporating it into the soil. The stubble inhibits weed establishment and reduces the need for herbicide sprays or cultivation. The barley is replanted after the berries are harvested.

Buckwheat is another traditional annual cover crop. It is well adapted to cool, moist climates, short growing seasons, and acid soils. Buckwheat is not winter hardy and is killed by mild freezing temperatures. Historically, buckwheat is considered to be effective at taking up large amounts of otherwise insoluble phosphorus, which becomes available to the fruit crop when the buckwheat crop is killed. Buckwheat can become a troublesome weed and should be tilled into the soil before the seeds ripen.

An effective mix for an annual cover crop might include 80 pounds (90 kg/ha) of wheat, barley, or oats, or 30 pounds (34 kg/ha) of common buckwheat mixed with four pounds (4.5 kg/ha) of clover or 100 pounds (112 kg/ha) of winter or spring peas per acre. A list of suggested cover crops is given in Table 9.4.

TABLE 9.4. Recommended cover crops for northern sites.

| Common name | Seeding rate lb/acre (kg/ha) | Establishment rate | Nitrogen requirement lb/acre (kg/ha) | Representative varieties |
|---|---|---|---|---|
| *Annual cover crops* | | | | |
| Barley (spring) | 120 (135) | fast | 30-50 (34-56) | Currently available agronomic varieties |
| Barley (winter) | 120 (135) | medium-fast | 30-50 (34-56) | Currently available agronomic varieties |
| Buckwheat (common) | 35-60 (39-68) | medium-fast | 10-20 (22-23) | Common Gray, Japanese, Silverhull |
| Buckwheat (tartary) | 25 (28) | medium-fast | 10-20 (11-23) | Common varieties |
| Oats (spring or winter) | 120 (135) | medium-fast | 30-50 (34-56) | Amity (W), Walken (W), Grey Winter (W), Cayuse (S), Border (S), Kanota (S) or other currently available agronomic varieties |
| Peas (winter or spring) | 120 (135) | medium-fast | by soil test | Austrian Winter (W or S), Melrose (S), Miranda (S), or other currently available agronomic varieties |
| Ryegrass (annual) | 30 (34) | fast | 30-50 (34-56) | Common variety |

| | | | | |
|---|---|---|---|---|
| Wheat (spring) | 120 (135) | 20-50 (23-56) | medium-fast | Currently available agronomic varieties |
| Wheat (winter) | 120 (135) | 30-50 (34-56) | medium | Currently available agronomic varieties |
| Grain/Pea | 80/100 (90/112) | 20-30 (23-34) | fast | Use varieties recommended in this table |
| *Permanent cover crops* | | | | |
| Fescue (hard) | 20 (23) | 20 (23) | slow | Durar |
| Fescue (sheep) | 20 (23) | 20 (23) | very slow | Covar |
| Fescue (tall) | 25 (28) | 20 (23) | medium | Alta, Fawn, Forager, Kenhy |
| Perennial rye | 25 (28) | 30 (34) | fast | Elka, Linn, Manawa (H1), Manhattan, Norlea, Pennfine |
| Russian wild rye | 30 (34) | 20 (23) | slow | Vinall, Swift, Cabree, Bozoisky |
| Siberian crested wheatgrass | 35 (39) | 20 (23) | medium | P-27 |
| Standard crested wheatgrass | 25 (28) | 20 (23) | medium | Nordan |
| White clover | 4 (4.5) | 0 | medium | Dwarf types |
| Strawberry clover | 4 (4.5) | 0 | medium | Common varieties |

*Note*: W = winter variety; S = spring variety.

## Herbicides

Herbicides are a valuable part of modern agriculture. When used properly, they provide excellent weed control with little or no damage to the environment. Herbicides are generally more cost-effective in controlling weeds than mulching, cultivation, and hand weeding because they decrease the labor needed to produce a crop. The disadvantages are that when they are used improperly, herbicides can be dangerous to field workers, damage the crop or other desirable plants, and disrupt the environment within the soil. Any time soils are kept vegetation free, whether with herbicides or other cultural practices, erosion and loss of soil structure become concerns. Some weeds also develop resistance to certain herbicides when the chemicals are used over a long period of time.

In Europe, large commercial farmers have largely replaced tillage with herbicides. In these systems, herbicides are applied to bare ground during the winter or early spring and prevent weed emergence the following summer. The main drawback with using herbicides for *Ribes* crops in North America is that few herbicides are registered for those crops. Because *Ribes* are minor crops, most pesticide manufacturers do not find it cost-effective to develop or register herbicides for them, and the situation is not likely to improve in the future. In 1991, 12 contact and preemergence herbicides were registered for gooseberry and currant production in the northwestern United States. By 2001, the number of available materials had decreased to six herbicides (William et al., 2001). Before using any herbicides or other pesticides, ensure that they are registered for your crop and location. Always follow label directions.

The two types of herbicides used in horticulture are contact and preemergence. The postemergence, contact herbicides glyphosate, paraquat, and sulfosate are registered for use on *Ribes* in the United States and are effective against many weeds. They kill *Ribes* along with weeds, however, and applicators must be very careful not to direct the sprays or allow them to drift against green stems or leaves.

Preemergence herbicides are generally applied to bare soil and prevent the germination and/or establishment of weeds. If they are registered for your area, preemergence herbicides simplify weed control and reduce the labor associated with hand weeding, cultivation, and mulches. Preemergence herbicides are usually applied to bare soil within alleys and/or crop rows during the early winter before the soil freezes or in the late winter or early spring before weed seeds begin to germinate. Some preemergence herbicides require irrigation or precipitation soon after application. Contact herbicides can be used to spot spray weeds during the growing season.

## PRUNING AND TRAINING

Pruning is necessary in *Ribes* to remove old, unproductive wood, ensure good light penetration (needed for proper bud development), facilitate pest and disease control, and ease picking. Pruning also stimulates growth and encourages the development of new replacement canes. When growth is weak, such as in older plants, a combination of pruning and fertilization helps to invigorate growth. Where growth is already vigorous, less severe pruning is needed.

Pruning is one of the tools used to train *Ribes* plants to desired shapes. Different training styles are used to meet different requirements. *Ribes* are usually trained to individual, freestanding bushes, although black currants are sometimes developed into hedgerows to reduce pruning labor. Where space is limited, any of these crops can be trained on trellises or as cordons. Gooseberries and currants may also be grafted to form "trees" (Card, 1907; Harmat et al., 1990).

Bush and hedgerow systems require less labor, maintenance, and skill than systems using trellises or cordons. Because they have many canes, bushes and hedgerows also have the advantage of remaining healthy if one or a few canes are lost to pests or diseases. When a borer or cane blight kills the trunk of a cordon or tree-trained plant, the entire plant can die unless one or more spur branches are left at the base of the plant to provide replacements. The negative feature of bushes is that canes may lie close to the ground where they interfere with cultivation and are exposed to high levels of moisture, pests, and disease organisms. Pruning helps maintain an open canopy to provide for good light and air penetration, and to reduce disease problems (Keeble and Rawes, 1948).

When pruning, use sharp tools to ensure clean cuts. High-quality, bypass hand shears and loppers generally give better results than anvil-type tools. Long-handled loppers and leather gloves are desirable when pruning gooseberries. Remove unwanted canes as close to the ground as possible, and remove drooping canes that lie close to the ground. Canes are normally not shortened or headed back unless they are damaged or diseased. When cutting back part of a cane, cut immediately above a side branch or strong bud. If at all possible, remove prunings from the field and burn them to kill pests and diseases. If the prunings cannot be removed from the field, use a flail mower, disk, or rototiller to chop up the canes. Be careful while pruning red currants, white currants, and gooseberries not to damage the spurs. Most of the fruit for these crops is borne on short spurs on two- and three-year-old canes.

Except for cordon-trained plants, most *Ribes* are pruned while they are dormant during the late winter and early spring, but they can be pruned any time after the leaves have dropped in the fall. In colder areas, wait until late winter or early spring to prune in order to identify and remove wood injured by winter stresses. Where winter injury is not a problem, fall pruning improves air circulation around dormant bushes during wet fall, winter, and spring months, and can decrease disease problems.

### Freestanding Bushes

#### Red and White Currants, Gooseberries, and Jostaberries

With mature red and white currant, gooseberry, and jostaberry bushes, the goal should be to keep three or four strong, new canes per plant each year, and to remove an equal number of the oldest canes. In this system, mature plants have nine to twelve canes after pruning, three to four each of one-, two-, and three-year-old wood. Remove all wood that is four years old or older (Strik and Bratsch, 1990). The following yearly schedule can be used to produce freestanding and trellised bushes:

Year 1      At the time of planting, remove all diseased or damaged canes, keeping as many as six or eight healthy, vigorous, one-year-old canes. Remove diseased or damaged canes and roots, and all woody roots that are kinked or which twist around the collar or back into the center of the plant. Make clean cuts and do not leave branch stubs.

Year 2      In early winter through early spring following planting, remove all but three or four of the most vigorous two-year-old canes from dormant plants. Leave four or five of the most vigorous one-year-old canes.

Year 3      During the second dormant season following planting, leave three or four canes each of one-, two-, and three-year-old wood.

Year 4      During the third and subsequent dormant seasons, keep three to four one-, two-, and three-year-old canes, and remove all canes that are four years old or older.

## Black Currants

Black currant plants are more vigorous than other currants and gooseberries, and more canes are retained during pruning. As a general rule, leave ten to twelve vigorous canes per bush. If the bushes are very vigorous, leave a few more canes. About half of the canes left after pruning should be one year old, with the remaining half being vigorous two-year-old canes that have an abundance of one-year-old shoots. Remove all canes that are more than two years old. Figure 9.1 illustrates pruning and training for untrellised bushes.

## Hedgerows

Rather than grow currants and gooseberries as individual bushes, some European growers develop hedgerows. Although the steps in pruning are similar to those for bushes, the plants in a hedgerow are set closer together in the rows, forcing the canes to "bow out of the row" (Harmat et al., 1990). Hedgerows are used primarily because they are better suited to mechanical harvesting than bush-trained plants (Nemethy, 1977). For hedgerows, plant the bushes about two to two-and-a-half feet (0.6 to 0.75 m) apart. Alternatively, unrooted cuttings can be stuck into the ground approximately one foot (0.3 m) apart. When pruning mechanically harvested bushes, it is important to keep flexible, erect canes and to remove any canes lying closer than 45° to the ground (Nemethy, 1977).

To simplify pruning and reduce labor costs, it is possible to use an alternate-year production method for black currants. In an alternate-year system, the plants are grown in a hedgerow as described previously or in a meadow system as described in Chapter 8. To begin an

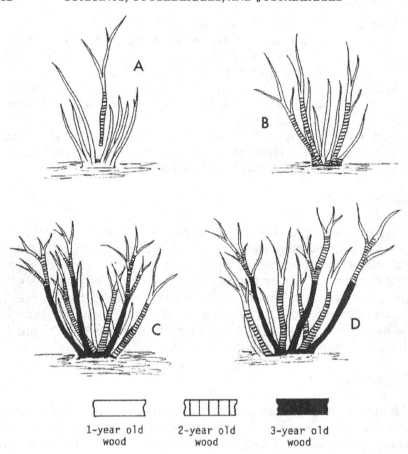

1-year old wood    2-year old wood    3-year old wood

FIGURE 9.1. Training red and white currants, gooseberries, and jostaberries as freestanding bushes. (A) At the time of planting, remove all two-year-old and older wood. Leave six to eight one-year-old canes. (B) During the first dormant season, remove all but three or four two-year-old canes. Leave four or five vigorous one-year-old canes. (C-D) During the second and subsequent dormant seasons, leave three or four each of one-, two-, and three-year-old canes. Remove all canes four years old or older.

alternate-year system, mow off established, dormant black currant canes about two inches (1 cm) above the ground. Allow all of the new canes that develop during the next spring to grow. During the following dormant season, remove only those canes that are diseased, damaged, or grow outside of the crop row. The crop is harvested during the following summer from second-season wood using handpicking, a conventional shaker-style mechanical harvester, or a combine-style mechanical harvester. If the canes are not cut off during harvest, mow off all of the canes one to two inches (3 to 5 cm) above the ground during the following dormant season. A hedgerow system for alternate-year production of black currants is illustrated in Figure 9.2.

## *Trellises*

Currant and gooseberry bushes can be trained to fan shapes that are supported by trellis wires. Their vigorous growth makes trellising jostaberries somewhat more difficult. Trellising conserves space, simplifies weed, pest, and disease control, and creates an attractive wall. This system can be used for commercial production, but it is best suited to hobby gardens (Harmat et al., 1990). Schuricht and

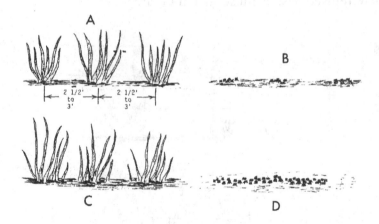

FIGURE 9.2. Hedge training for alternate-year production of black currants. (A) Plant bushes two feet apart in rows. Grow the plants for several years as free-standing bushes to allow them to become fully established before forming hedgerows. (B) During the first dormant season of hedgerow training, mow off all canes one to two inches (3 to 5 cm) above the ground. Trim prunings to six-inch (15 cm) lengths and plant every 12 inches (30 cm) within the rows between the original bushes. (C) During the second season, and every other year after that, keep rows narrowed to about one foot (30 cm) for hand harvesting or a width suitable for the mechanical harvester used. Do not thin canes within rows. Follow standard production practices and harvest. (D) During the third dormant season, and every other year after that, mow off all canes about two inches (1 cm) above the ground.

Schwope (1969, p. 497) compared conventional bush culture with production of red currants on wire frames and concluded that the wire support systems they investigated "cannot be recommended for practical growing."

Start a trellis system by setting individual gooseberry and black, red, and white currant plants every three to four feet (1 to 1.3 m) next to a trellis that has three to five horizontal wires spaced about one foot (30 cm) apart. Space jostaberries four to five feet (1.3 to 1.6 m) apart. The height of the trellis depends on the vigor of the bush. Currants and gooseberries can be trained on smaller trellises than jostaberries. As side branches develop on the canes, remove long, thin branches and keep shorter, stronger branches (Harmat et al., 1990). Tie the branches to the trellis wires, keeping the ties loose and checking them regularly to ensure that they do not cut into or girdle the branches. When pruning, follow the guidelines previously given for bushes. A typical trellised bush is illustrated in Figure 9.3.

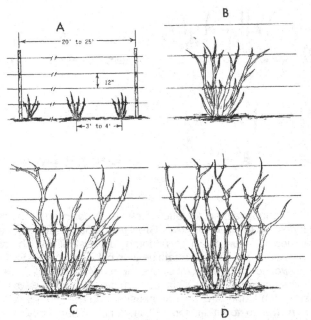

FIGURE 9.3. Trellising for currants and gooseberries. (A) Plant bushes three to four feet (1 to 1.2 m) apart. Tie the canes to the bottom wire. (B-D) As side branches develop, tie them to the trellis wires. Prune using the methods given for freestanding bushes.

## *Cordons*

Cordon-trained plants have from one to about three straight trunks approximately six feet (2 m) tall that are supported on stakes. Any currant, gooseberry, or jostaberry can be trained to a cordon. As with trellises, cordon-trained plants take up less space than bushes, are attractive, simplify picking, improve spray penetration, increase air and light penetration, and can increase gooseberry fruit size. Cordon training, however, is labor intensive, reduces fruit yields, and can increase sunscald and berry cracking on sunny sites. Another problem with cordons is that damage to the trunk can kill an entire plant. For these reasons, cordons have generally been used only in hobby gardens. According to S. A. McKay (personal communication), however, some commercial growers in Holland have adopted the use of cordons in highly managed systems to produce red currants and gooseberries for fresh markets.

The following procedures apply to traditional English cordons that have a single trunk for both currants and gooseberries. Red currant cordons in Holland may have up to three trunks, and both currant and gooseberry cordons have an extra, short cane at the base of each plant to replace damaged trunks. Otherwise, development and management of the cordons are generally similar with the following exceptions. In England, a cordon is often developed by cutting all canes off near the ground and training up a new cane to serve as the trunk. In Holland, one medium-sized cane (gooseberry) or three medium-sized canes (red currant) are selected and serve as the trunk(s), reducing establishment time by one to two years. English cordons use fruiting spurs that are kept for several years before replacement. In the system described by McKay (personal communication) one-year-old fruiting canes are removed immediately after harvest, along with small and crowding branches. This strategy leaves medium-sized branches that will bear the following year.

To establish a cordon system, set out individual plants in rows at least six feet (2 m) apart. When the plants to be cordoned are especially vigorous (black currants and jostaberries), space the rows farther apart. Within the rows, set gooseberries, red currants, and white currants two to three feet (0.6 to 1 m) apart. Black currants and jostaberries should be set about four feet (1.3 m) apart.

Place stakes that are at least eight feet (2.6 m) long on the north sides of the plants. The stakes should be set at least two feet (0.6 m) deep into the ground. Steel fence posts make excellent stakes and 3/4-inch to one-inch diameter (1.9 to 2.5 cm) galvanized metal, electrical conduit has also been used effectively. While the plant is still dormant, select a single, straight, vigorous shoot to form the trunk and remove all other shoots. One-year-old canes probably give the best results. If no suitable shoot is available for the trunk, cut all growth back to two or three buds to force new growth. From that new growth select a straight shoot to form the trunk.

After a shoot is selected, remove about one-half of it, cutting it back to a point just above a strong bud. Cut any laterals on the retained shoot to a single bud. Remove all suckers at the ground as soon as they develop. As the trunk-to-be grows, tie it to the stake, taking care not to allow the ties to cut into or girdle the shoot. When it has grown to a height of 18 inches (45 cm), remove all laterals within about six inches (15 cm) of the ground.

During the first few dormant seasons, cut the central leader to retain one-half of the previous season's growth. Make the cut about one-quarter of an inch above a strong bud. Remove any poorly positioned or crowded laterals and prune the remaining laterals to about three buds. For most gooseberries, cut the laterals to upward and outward pointing buds. For red, white, and black currants, jostaberries, and vigorous gooseberry cultivars, cut the laterals back to downward and outward facing buds. Pruning in this fashion helps to develop horizontal branches.

Mature cordons are pruned during the summer. During June or early July, remove crowded laterals, spacing laterals and fruiting spurs two to three inches (5 to 7 cm) apart on the central leader. Remove laterals that are four years or more old. After growth has slowed or stopped in late July or early August, shorten the laterals to about five leaves on the current season's growth. During the first few seasons, do not summer prune the central leader. After a strong central leader has developed and reached a height of about six to seven feet (2 m), shorten the leader to a single bud on the current season's growth during summer pruning in July or August. Using this method, it should be possible to keep the trunk about six to seven feet tall for many years. Figure 9.4 illustrates the steps involved in developing a cordon.

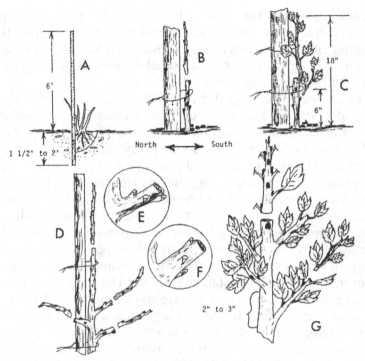

FIGURE 9.4. Training currants, gooseberries, and jostaberries to cordons. (A) Plant bushes on the south sides of stakes. (B) Select a straight, upright, vigorous, one-year-old cane. Cut off all other canes at the ground. Cut off about one-half of the selected cane, making the cut just above a strong bud. (C) When the central leader is about 18 inches (45 cm) tall, remove all side branches within six inches (15 cm) of the ground. (D) During the dormant seasons, cut off about one-half of the central leader shoot that formed during the preceding growing season. Shorten new lateral shoots to about three buds each. (E) For currants, jostaberries, and vigorous gooseberries, cut to a downward-pointing bud. (F) For less vigorous gooseberries, cut to an upward-pointing bud. (G) During July or August, pinch lateral shoots to about five leaves each. Begin summer pruning the central leader when it grows to about six feet (2 m) tall. When the central leader reaches the desired height, pinch off all but one bud on the current season's shoot of the central leader during summer pruning (Eppler, 1989).

## *Gooseberry Trees*

Training gooseberries to "tree" forms is similar to using a cordon and is used to raise the canes above the ground. Raising the canopy on a clean trunk facilitates weed control, reduces the incidence of mildew, and is adapted to mechanical harvesting. The method was popular in the United States during the early days of *Ribes* culture. By the beginning of the twentieth century, however, it had largely fallen out of favor and the "trees" were considered oddities, suited only as curiosities for home gardens (Card, 1907).

Tree culture is the most difficult of the five systems described in this guide, with the possible exception of cordons, and requires the greatest investment of labor and materials. While gooseberry trees can be developed on their own roots, many are created using scions grafted onto rootstocks. In the case of a grafted tree, ailing or aging plants cannot be produced using suckers from the collar, since these develop from the rootstock and do not produce the desired fruit. For that reason, damaged or aging trees must be replaced using new, grafted plants. Despite its disadvantages, however, gooseberry tree systems have been used commercially in Europe since the 1920s, primarily on small farms in Hungary. Most commercial growers today, however, prefer to use freestanding bushes. The steps in developing a grafted gooseberry or currant tree are shown in Figure 9.5.

The simplest method for developing a tree is to start with a single, vigorous cane. Rub off the buds or prune off all branches less than 12 to 18 inches (30 to 45 cm) from the ground. Select six or eight shoots to serve as the branches of the tree and cut them back to lengths of four to six inches (10 to 15 cm) each year until the tree is fully grown. This practice is similar to pruning a cordon, as described earlier. All lateral shoots that develop from the branches are cut to within an inch of the branch each year (Card, 1907).

Grafted trees are formed by grafting a scion from a desired cultivar onto a vigorous rootstock. Recommended rootstocks are *Ribes aureum* clones 'Brecht' and 'Pallagi 2', neither of which is readily available in the United States at the present time. These clones have the advantages of good graft compatibility, vigorous growth, limited tendency to sucker, and are relatively resistant to insects and diseases. Seedlings of *R. aureum* may also serve as acceptable rootstocks.

Rootstocks are grown from mother plants in stool beds. The mother plant is cut low to the ground and allowed to develop shoots. After the shoots have grown about one foot (30 cm) tall, mineral soil, potting soil, or sawdust is mounded around the base of the shoots to a depth of six inches (15 cm) and kept moist. Roots form on the base of the shoots. Grafting takes place during the growing season after

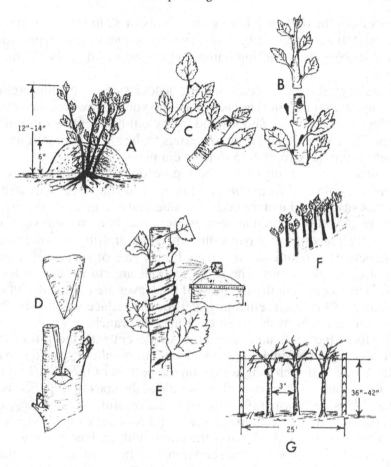

FIGURE 9.5. Training gooseberries to tree forms. (A) Develop rootstocks *(Ribes aureum)* in a stool bed. (B) Two days before grafting, cut shoot tips off rootstocks. Strip off the top three leaves and remove the second bud from the top of rootstocks. (C) During July or August, on the day the grafting is to take place, collect two- to four-inch-long (5 to 10 cm) scion cuttings from healthy, current-season growth or one-year-old wood with a side branch. Scion shoots must have well-developed buds. (D) At the time of grafting, make a one-half-inch deep (1 cm) cut in the top of the rootstock, parallel to the top bud. Make a matching wedge on the bottom of the scion cutting and insert the scion into the rootstock. (E) Fasten the graft union together with a banding rubber or plastic tape. If banding rubbers are used, make the graft union airtight using tar-based pruning compound. (F) Dig dormant, grafted, rooted layers from the stool bed and plant in nursery rows for one year. (G) Plant trees and support on trellis wires. Ensure that ties do not cut into the trunks.

shoots from the rootstock have grown to about 32 to 40 inches (80 to 100 cm) tall. Green grafting, using a whip and tongue union, provides better success than grafting hardwood cuttings during the dormant season.

For the graft to be successful, the rootstock should be undifferentiated (not woody). Using shoots that are too young, however, is not advisable because young shoots tend to dry out before the graft union heals. The rootstock and scion materials at the site of the graft union should be about 3/16 inch (5 mm) in diameter.

Collect scion cuttings from the top portions of mother plants. If scions are collected from current season's growth, select only material that has formed mature buds. Commercial nurseries prefer to use semi-woody shoot tips that have set terminal buds or one-year-old side branch sections with two to three buds for grafting onto the green rootstock. The mature scions give a higher rate of successful grafts than do younger scions. The scion sections are cut to two- to four-inch (5 to 10 cm) lengths and most of the leaves are removed. Cut off the outer half of each remaining leaf to help reduce water loss. Remove all leaves from scions taken from side branches.

To form the graft union, make a single cut into the top of the rootstock, one-half inch (13 mm) deep and parallel to the top bud. Make a one-half-inch long wedge on the bottom of the scion and insert the scion into the rootstock. Ensure that the apex of the scion is at the top. Upside-down grafts will not be successful. Use grafting tape or a banding rubber to hold the scion and rootstock together. When banding rubbers are used, cover the union with tar-based tree wound compound to exclude air and keep moisture in. In Europe, one-half-inch long sections of 3/16-inch-diameter (5 mm) PVC tubing are used to hold graft unions together. Insert the tube over the rootstock before grafting. Once the union is made, pull the tube up and over it to hold the scion to the rootstock. As an alternative to whip grafting, you may also choose to side graft the scion onto the rootstock in August.

The grafted rootstock is dug from the stool bed in late fall. It can then either be planted directly into a nursery or placed into cold storage during the winter. Ensure that the roots do not dry out during storage. If the stock is planted in the fall, place about four inches (10 cm) of mulch around it to reduce frost heaving.

When planting into a nursery bed, set the rootstocks about four to six inches (10 to 15 cm) apart in double rows eight inches (20 cm) apart. Pairs of double rows should be spaced four feet (1.3 m) apart. Irrigate the planting beds regularly during the growing season. After budbreak, gradually remove developing shoots on the rootstock when they are one-half to three-fourths-inch (2 to 3 cm) long. Use a knife to cut the shoots off cleanly at the rootstock. Do not remove all of the shoots at once; doing so weakens the rootstock and slows root development. Remove the grafting tape, banding rubber, or PVC tube at the same time the rootstock shoots are removed.

Keep the grafted trees in the nursery until they have become dormant in the fall, and then dig and transplant them. Before planting, prune off all but three side branches, spacing them one and one-half to two inches (4 to 5 cm) apart. Planting steps are the same as those described in Chapter 8.

Support the trees on a single-wire trellis, with the wire about six inches (15 cm) above the head of the tree. Use ten to twelve gauge wire and support it on the top of wooden posts. Use a two- to four-inch (5 to 10 cm) diameter post for support about every 25 feet (8 m) and ensure that the ends of the trellises are solidly anchored. Attach the trees to the trellis wire with string or other fasteners. Be careful not to allow ties to girdle the trees. Maintenance of gooseberry trees is the same as that described previously for bushes.

## *POLLINATION*

Various wild and domestic bee species and other insects pollinate *Ribes*. Being among the earliest blooming fruits, these crops may suffer from poor pollination during cold, wet springs when bee activity is low. As discussed in Chapter 2, while most red and white currant and gooseberry cultivars are self-fertile, the degree of self-fertility varies according to the climate. Fruit set and size may be improved by including at least 1 to 2 percent of the bushes as pollinizing cultivars. Partial or total self-sterility can also be a problem in black currants and jostaberries. Selection of pollinizing cultivars is discussed in Chapter 2 and planting designs to enhance cross-pollination are described in Chapter 5.

To ensure adequate pollination, place one or two bee hives per acre (2 to 4 hives/ha) when about 25 percent of the flower clusters have opened. During years when weather during the bloom period is cool, wet, or windy, use more hives. Locate hives in the centers of fields, rather than at the edges. Place the hive fronts facing south, and elevate the hives on supports, if necessary, to provide good light exposure and air movement. Before bringing hives into the field, eliminate all weed and groundcover flowers by mowing. Bees are strongly attracted to wildflowers, such as dandelions, and often prefer them to *Ribes* flowers. Place pans of water near hives and insert flat sticks into the containers to allow bees that fall into the water to climb out.

Do not spray pesticides that are harmful to bees during the bloom period, and ensure that harmful pesticides that have long residual activity have not been sprayed in or near the plantation prior to bloom. Microencapsulated methyl parathion is especially toxic to bees. If pesticide applications are necessary during the bloom period, apply the sprays in the late evening, after bee activity has ceased for the day. Pesticide residues can still kill or injure bees during the following days, depending on the pesticide used.

## WINTER PROTECTION

As discussed earlier, *Ribes* are among the most cold-hardy domestic fruits and normally perform best in the cooler portions of North America. In most cases, keeping the plants healthy through proper fertilization, weed and pest control, disease control, pruning, and irrigation are the only steps needed to ensure good winter survival. In very cold, northern regions, apply soil or an organic mulch around the bushes in late fall or early winter after the canes are dormant to help protect the roots and collars from freezing. This practice should ensure new cane production even if the old canes are occasionally killed. Protecting the plants from drying winter winds also helps ensure winter survival, and windbreaks may be beneficial on windy sites. A calendar of typical yearly activities is given in Table 9.5. Dates in the table apply to cool, short-season sites. In warmer areas, cultural practices begin earlier in the season and may extend longer.

TABLE 9.5. Annual calendar for managing established currant, gooseberry, and jostaberry crops.

| Month | Procedure |
| --- | --- |
| January–February | Arrange for beehive rentals during April through May. |
| | Order harvest baskets and flats, pesticides, fertilizers, and other supplies. |
| | Arrange for contract labor, if required. |
| March | Apply preemergence herbicides when ground thaws, if required. |
| | Test and service refrigeration equipment. |
| | Every two to three years, test soil samples from each field for pH and organic matter. |
| April | Prune. Remove prunings from fields and burn, if possible. |
| | Apply lime sulfur, as well as dormant oil sprays just before budbreak. |
| | Apply slow-release fertilizers within berry rows. Amounts of fertilizer are listed in Table 9.3. |
| | Inspect and repair fences, irrigation systems, and other facilities and equipment. |
| | When 25% of flower clusters are in bloom, place hives in fields. Use one or two hives per acre. |
| May | Apply 1/3 of quick-acting nitrogen fertilizers, if used, to fruit and cover crops. |
| | Replenish mulches in crop rows, if necessary. Continue rodent control program. |
| | Mow annual cover crops and leave stubble. |
| | Mow permanent cover crops or cultivate alleys, as needed. |
| | Cultivate, hand weed, and/or spot spray as needed. |
| | Inspect weekly for insect, mite, disease, bird, and gopher problems. Control, as needed. |
| | Irrigate, as needed. |
| June | Apply 1/3 of quick-acting nitrogen fertilizers to fruit and cover crops. |
| | Mow permanent cover crops or cultivate alleys, as needed. |
| | Spot spray or hand weed within rows. |
| | Inspect weekly for insect, mite, disease, bird, and gopher problems. Control, as needed. |
| | Irrigate, as needed. |

TABLE 9.5 *(continued)*

| Month | Procedure |
|---|---|
| July | Apply 1/3 of quick-acting nitrogen fertilizers to fruit and cover crops. Use the same amounts as for May application. |
| | Begin harvest. |
| | Mow permanent cover crops or cultivate alleys, as needed. |
| | Spot spray or hand weed within rows. |
| | Inspect weekly for insect, mite, disease, bird, and gopher problems. Control, as needed. |
| | Irrigate, as needed. |
| August | Continue harvest, if needed. |
| | Collect leaves during the first week of the month and send to laboratory for foliar test. |
| | Plant annual cover crops. |
| | Mow permanent cover crops or cultivate alleys, as needed. |
| | Spot spray or hand weed within rows. |
| | Inspect weekly for insect, mite, disease, bird, and gopher problems. Control, as needed. |
| | Irrigate, as needed. |
| September | Mow cover crops or cultivate alleys, as needed. |
| | Spot spray or hand weed within rows. Inspect weekly for insect, mite, disease, and gopher problems. Control, as needed. |
| | Irrigate, as needed. |
| | Drain and winterize irrigation system. |
| October | Service rodent bait stations in mulched fields. |
| | Evaluate production practices and modify, if needed, for next year. |
| November | Apply preemergence herbicides before ground freezes, if required. |
| | Apply Bordeaux spray to dormant bushes after leaf fall. |

# Chapter 10

# Pests and Diseases of *Ribes*

American powdery mildew, white pine blister rust, leaf spot diseases, imported currant worm, and gooseberry maggot are common problems faced by *Ribes* growers in North America. Other pests and diseases can become serious in some regions. When native or cultivated *Ribes* are located nearby and as planting sizes increase, pest and disease pressures often increase as well. Successful currant, gooseberry, and jostaberry operations require aggressive pest- and disease-control programs.

Purchasing high-quality, disease- and pest-free stock is critical in developing a productive planting. Deal only with certified nurseries experienced with *Ribes* production. Obtaining plants from commercial fields or hobby gardens greatly increases the risk of introducing pests and diseases into your planting.

After planting, the most important step in controlling pest and disease problems is to scout the fields thoroughly and regularly. Learn to recognize and control potential pests and diseases, and know when and where to look for them. Inspect all fields at least weekly from early spring though fall. Keep records of all insect, mite, pest, and disease problems. Use records from past years to anticipate potential outbreaks. Also keep records of beneficial insects and mites and learn how to manage their populations.

Use all available methods of pest and disease control. Good sanitation practices, such as burning prunings, greatly reduce pest and disease problems. Controlling weeds and insects in adjacent areas also helps reduce pest and weed pressures in fields. Proper pruning helps maintain open canopies and reduces diseases and pests by increasing air circulation and improving spray penetration.

Relatively few pesticides are registered in North America to control pests and diseases on *Ribes* crops. In cases where pesticides are available, use them only when necessary to avoid economic losses.

Excessive or improperly timed sprays kill beneficial insects and mites that help to control pests. Spider mite outbreaks can often be traced to improper pesticide use. A grower's goal should be to keep pest populations below economic levels, not necessarily to eradicate the pests.

If they are registered for your area, dormant sprays of lime sulfur, Bordeaux mixes, and crop oils help reduce pest and disease problems. As a general rule, apply lime sulfur or Bordeaux sprays in the fall or early winter after leaf drop to control fungal diseases. As the buds begin swelling in the spring, apply another sulfur spray to further reduce powdery mildew and other fungi. Note that Bordeaux mixes contain copper, which can accumulate in the soil if applied incorrectly. A good practice is to limit Bordeaux sprays to once each year, preferably the fall, and to keep the spray on the bushes, rather than on the ground.

Applying dormant oil sprays when *Ribes* buds begin swelling in the spring helps control mites, aphids, and other insect pests. Some commercially available sulfur products can be tank mixed with and applied at the same time as dormant oil. Consult with your state or provincial agriculture extension specialists and commercial pesticide suppliers to identify materials that are registered and available in your area. Whatever dormant spray materials you use, confine the spray to the canes and mix up only as much spray material as will be used that day to avoid disposal problems. Follow label directions carefully when using any pesticide.

## *PESTS OTHER THAN INSECTS AND MITES*

### *Birds*

Birds can seriously damage *Ribes* crops. Covering the bushes with commercially available bird netting provides the most effective control. Ensure there are no large holes in the net and that it drapes all the way to the ground. Staking nets down at their bottom edges is an advisable practice, especially in windy areas.

## Nematodes

These microscopic worms can cause problems in *Ribes* crops. The most serious nematode for *Ribes* growers in North America is the American nematode, *Xiphinema americanum,* which appears to spread currant mosaic virus. Before planting, have your soil tested to identify nematode pests present and take steps to control them. Fumigation is used to control nematodes in commercial fields. Pesticides called nematicides are available to control nematodes. Some *Brassica* (mustard, canola, and radish) varieties produce chemicals toxic to nematodes and can prove useful in reducing pest populations when grown and tilled in before planting. At the time of this writing, however, none had been tested against the American nematode.

Plant only clean, healthy stock that you are confident is free of viruses. Rogue out and dispose of infected plants.

## INSECTS, MITES, AND INVERTEBRATE PESTS

Because pesticide registrations change constantly and vary from region to region, no specific chemical recommendations are given in the following sections. If a pest or disease can be controlled with pesticides, a note to that effect is included in the description. Check with your agricultural chemical supplier to identify materials registered for particular pests in your area.

### Cottony Maple Scale (Pulvinaria innumerabilis)

Immature scales are flattened and brown to yellowish-green in color. Conspicuous cottony egg sacs are two to three times the length of the scales' bodies. The insects feed on foliage and reduce plant vigor, causing twig dieback. Control with crop oils applied to dormant plants during the early season before buds open.

### Currant Aphid (Cryptomyzus ribis)

Currant aphids are common on red and white currants and occasionally found on gooseberries. The insect overwinters in the egg stage on the bark of new canes. The small, yellowish aphids begin to appear when leaf buds open in the spring. They feed on the undersides of foliage, causing the leaves to redden and assume a distorted, cup shape. Honeydew accumulation on foliage and fruit is unsightly and makes fruit unmarketable. Winged aphids develop in early summer and fly to other non-*Ribes* hosts. The aphids migrate back to the *Ribes* in the fall to mate and lay eggs. Pesticides are available to control this pest.

### Currant Borer or Clearwing Moth (Synanthedon tipuliformis)

Adults are clear-winged, blue-black, wasplike moths with yellow markings. They appear in late May or early June and deposit eggs on canes. The pale-yellow larvae tunnel in the canes, weakening the plants and causing yellowing of the foliage and leaf wilting during summer and autumn. Larvae overwinter in the tunnels. Red currants are the most susceptible host, but this pest can attack any currant or gooseberry. To control, prune out and burn all affected shoots. Check state or provincial rules for registered pesticides.

### Currant Fruit Fly or Gooseberry Maggot (Epochra canadensis)

During the summer, larvae bore out of infested berries, fall to the ground, and enter the soil. There they overwinter as pupae in brown cases about the size of wheat grains. Flies emerge from the soil in the spring and soon lay eggs in developing gooseberry or red, white, or black currant fruit. Susceptibility of jostaberries is not yet documented. Adult flies frequently rest on brush and trees adjacent to *Ribes* plantings. The currant fruit fly is one of the most serious pests of gooseberries and currants. Shallow cultivation under bushes and in alleys can help to expose egg cases and larvae to predators. Plastic mulches in the berry rows can help prevent the larvae from entering the soil. At one time, chickens, which scratch up and eat the larvae, were placed in *Ribes* plantings to control the fruit fly. Pesticides are available to control currant fruit fly.

## *Currant Stem Girdler* (**Janus integer**)

The currant stem girdler is a sawfly that emerges in late April or early May. Oviposition injury reduces cane length. Check state or provincial rules for registered pesticides.

## *Eriophyid Mites*

These various species of microscopic mites feed on the buds and foliage of *Ribes*. Up to 30,000 mites can infest a single bud on black currant. There appear to be 14 species of eriophyid mites that attack currants and gooseberries throughout the world, but only one or two of the species occur in North America. The mites feed on or within buds and on the leaves, causing misshapen leaves, cavities at the bases of veins on the bottoms of the leaves, or small blisters on the leaves. Bud-feeding species can cause buds to either shrivel or become enlarged (Amrine, 1992). The most serious of the mites is the black currant gall mite or big bud mite (*Cecidophyopsis ribis* Westw.), which is the vector for reversion disease in black currants. The only effective control measures are to plant healthy stock and rogue out infested bushes. The black currant gall mite is not present in North America, and no chemicals are registered in the United States to control this pest.

## *Fourlined Plant Bug* (**Poecilocapsus lineatus**)

The nymphs of this North American native are small and can be recognized by the shining, vermilion-red color of the body and blackish spots on the thorax. The adults, which are about three-tenths of an inch (8 mm) long, are bright yellowish-orange and have four black stripes along their backs. The pest appears about May on the tips of succulent shoots and acts by sucking sap out of the leaves. The feeding site develops into a small dark spot and may fall out. Heavy feeding can kill leaves and shoots. Since the insect overwinters as eggs on the tips of the shoots, dormant oil sprays are among the most effective controls.

### Gooseberry Cambium Miner (Opostega scioterma)

Adults are gray moths, one-fourth inch (6 mm) long, which appear in June. Slender, semitransparent, downworm-like larvae mine up and down the cane cambium, stunting or killing the tips of new canes. No insecticides are registered for this pest. Prune out and burn infested canes.

### Gooseberry Fruitworm (Zophodia convolutella)

This pest attacks the fruits of gooseberries, currants, and probably jostaberries by boring into the berries. One larva can destroy several berries. Larvae are green with a yellow tinge and darker stripes on their sides. The pests overwinter as pupae underground or beneath litter, emerging as adult moths in the spring. The adults have ash-colored wings with dark markings and a wingspan of nearly one inch (25 mm). The moths lay their eggs in the *Ribes* flowers. No pesticides are registered for this pest, but programs to control currant fruit fly and tent caterpillars should also control gooseberry fruitworm. Shallow mechanical cultivation under the bushes can help expose and kill pupae.

### Imported Currantworm (Nematus ribesii)

Full-grown larvae are about one-half inch (12 mm) long, greenish in color, and often have dark body spots, especially when partially grown. The larvae feed along leaf margins and can quickly defoliate plants when insects are numerous. Adults are black sawflies with yellowish abdomens. Cultivars vary somewhat in their susceptibility, but all currant, gooseberry, and jostaberry cultivars appear susceptible. The larvae are easily controlled with pesticides, but with as many as three generations of the pest per year, heavy infestations can develop rapidly. Regular field scouting and prompt action are necessary to prevent bushes from being seriously damaged. Although the larval worms resemble lepidopterous caterpillars, they are not related and *Bacillus thuringiensis* products will not control imported currant worm.

## Pacific Flat-Headed Borer or Flat-Headed Borer (Chrysobothris mali)

The larvae attack the roots and canes of all *Ribes,* as well as many native plant species. Stressed plants are particularly susceptible to attack. Adults have flattened, elongated-oval, brownish bodies one-fourth to one-half inch (6 to 12 mm) long. The larvae are yellowish-white and slender, with the front part of the body broadly expanded and flattened. Eggs are deposited in June or July in cracks on the bark of trees and shrubs. The eggs hatch from mid-June to mid-August. The larvae attack both outer and inner bark, plugging their tunnel with frass as they feed. The borer overwinters in the trunk as a larva, emerging as an adult in the spring. Damage first appears as a wet spot on the bark. The bark covering the wound soon becomes roughened, cracks, and falls off. No chemicals are registered for control of this pest on *Ribes.* Prune out and burn infested canes.

## San Jose Scale (Quadraspidiotus perniciosus)

Adult scales are nearly circular and slightly convex. The color is dark or blackish when small but gray when fully developed. The scales attack the leaves, flowers, shoots, and canes of all *Ribes.* Control with crop oils and lime sulfur. Spray to thoroughly cover shoots while plants are dormant in early spring before buds break.

## Slugs

The various genera and species of slugs attack the leaves and fruits of all *Ribes.* Slugs are active at night and during cool, wet weather. Overhead irrigation creates an ideal environment for slugs, which hide during the day in cracks in the soil and under debris. Silvery-colored slime trails on damaged plant parts are evidence of their presence. Grass, weeds, and debris within fields provide habitats for slugs, and good sanitation and weed control help control them. When using overhead irrigation, irrigate early in the morning to allow foliage to dry before night. Prune bushes to open shapes to facilitate drying and air movement. Slugs can infest perennial legumes. In areas where slugs are a problem, avoid perennial clovers in cover crops and rotate out of alfalfa and other perennial legumes for one year prior to planting *Ribes*. Pesticides and poison baits are commercially available. Baits applied before fall rains will kill slugs before they can lay eggs. A home remedy for slugs is to make traps by placing shallow containers filled with beer in a garden. Slugs are attracted to the odor of brewer's yeast and drown after crawling into the containers.

### Tent Caterpillars *(**Malacosoma** spp.)*

Larvae are large brown, blue, or orange hairy caterpillars that live in silken tents. They feed on leaves and are often found together in large numbers. *Bacillus thuringiensis* (BT) bacterial insecticides are effective against these pests. Other pesticides are also available.

### Twospotted Spider Mite **(Tetranychus urticae)**

Mites overwinter as adults on weeds and debris at the bases of host plants. They are about one-fiftieth of an inch (0.5 mm) long, have eight legs, and are light tan or greenish in color, with a dark spot on each side. Feeding reduces plant vigor and may cause leaves to turn brown and drop. Spider mites become especially troublesome when predator mites and insects are killed by improper pesticide use. Maintaining a cover crop between berry rows can provide habitat for predator mites and assist in spider mite control. Pesticides are available to control spider mites.

## DISEASES

Parts of the following disease descriptions are adapted from the *2001 Pacific Northwest Plant Disease Management Handbook* (Pscheidt and Ocamb, 2001) and from *Diseases of Fruit Crops* (Anderson, 1956).

### Angular Leaf Spot **(Cercospora angulata)**

This fungus affects the leaves of all *Ribes* and can cause serious damage. Large brown or gray spots develop on the leaves. Black specks visible in the centers of the spots are fruiting bodies. One control measure is to prune bushes to open shapes to improve air circulation. Avoid crowding bushes by planting too closely together. If overhead irrigation is used, irrigate early in the morning to allow foliage to dry during the day. Bordeaux and lime-sulfur sprays applied during the growing season also give good control. Be careful when using sulfur sprays on nondormant bushes, as they can damage actively growing tissues during hot weather.

### Anthracnose Leafspot *[*Drepanopeziza ribis *(Kleb.)* von Hohn]

This fungus overwinters on dead leaves and is especially serious during wet seasons. It affects currants and gooseberries. Symptoms include small, dark brown, round to irregular leaf spots that, when abundant, can cause the leaves to yellow and drop by midseason. Leaf spots may contain small, grayish fruiting bodies. On currants, the fruits, as well as the leaves, may show spotting resembling fly specks. Severely infected berries crack open and drop. Anthracnose reduces plant vitality, growth, and productivity. Control by removing dead leaves or cultivate them into the soil before new leaves develop. Prune to open bush shapes to provide good air circulation. Anthracnose control can be combined with powdery mildew control programs. Chemical controls are also available.

### Armillaria Root Rot (Armillaria mellea)

This fungus is sometimes called the oak root or shoestring fungus and is often present on newly cleared forest and woodland sites. *Armillaria* frequently kills oaks, pines, willows, fruit and nut trees, caneberries, grapes, roses, and many other perennial woody plants. The fungus attacks the canes and roots of all *Ribes*. Infected plants decline, produce no new growth, and gradually die. During autumn, honey-colored mushrooms or toadstools often appear at the bases of infected plants. When dug up, black shoestring-like strands of the fungus "roots" (rhizomorphs) can be found clinging to the roots of diseased plants, together with white felts or mats of fungus growth. The fungus grows, for the most part, between the bark and the wood of the roots and can survive on dead stumps and cane stubs for years. Control by removing and destroying diseased plants. Do not replant for two or three years. If *Armillaria* is present and you must plant, fumigate the soil.

## White Pine Blister Rust (**Cronartium ribicola** *Fisch.*)

This fungus attacks both wild and cultivated *Ribes*. European black currants *(Ribes nigrum)* are more susceptible than red currants, white currants, or gooseberries, although at least some cultivars in all domestic *Ribes* crops appear susceptible. Uredia can form on jostaberry leaves under conditions of high inoculation. In most years jostaberries remained uninfected under natural conditions in tests conducted in Corvallis, Oregon. The black currant cultivars 'Coronet', 'Crusader', 'Consort', and 'Titania' have been reported to be immune to blister rust, although 'Titania' developed blister rust symptoms in one Danish trial. Additional black currant cultivars immune or highly resistant to blister rust are listed in Chapter 7, and Table 12.2 lists black, red, and white currant cultivars immune or highly resistant to the disease. The wild species of stink currant *(R. bracteosum)*, flowering currant *(R. sanguineum)*, Sierra gooseberry *(R. roezlii)*, and Sierra currant *(R. nevadense)* are highly susceptible. Five-needled pines are alternate hosts for the fungus. Rust on white pine has caused severe losses of timber trees. Plant susceptible currants and gooseberries at least one mile from five-needled pine trees. Better yet, plant only rust-resistant or immune cultivars in regions where susceptible pines are found. Susceptible pines in North America include whitebark pine *(Pinus albicaulis)*, limber pine *(P. flexilus)*, eastern white pine *(P. strobus)*, western white pine *(P. monticola)*, sugar pine *(P. lambertiana)*, bristlecone pine *(P. aristata)*, foxtail pine *(P. balfouriana)*, and torrey pine *(P. torreyana)*.

Small cuplike spots formed from minute, orange, hairlike structures develop on the undersides of affected currant and gooseberry leaves. Spores from these cups cause additional infections of *Ribes* leaves throughout the summer. Affected plants are weakened and may show premature defoliation. Spores from infected *Ribes* are heavy and travel less than a mile unless carried on strong updrafts along mountain slopes. Spores from infected pines can travel for many miles to infect *Ribes* but cannot infect other pines. No chemicals are registered to control blister rust, but Picton and Hummer (2003) found that horticultural-grade mineral oil at a rate of 8 ml/L in water and applied to run-off every two weeks reduced blister rust symptoms in three highly susceptible black currant cultivars. The oil treatments did not reduce blister rust symptoms in four more resistant cultivars. Where horticultural mineral oil is used during the growing season in a powdery mildew control program, some relief from blister rust symptoms may be obtained for highly susceptible black currants. Regardless of the limited success oil applications provide in controlling blister rust, the authors strongly recommend planting only rust-resistant currant and gooseberry cultivars in regions where susceptible pines are found.

### *Botrytis Dieback and Fruit Rot* (**Botrytis cinerea** *Pers.*)

This fungus attacks the canes, leaves, flowers, and fruits on all currants and gooseberries. The disease is especially prevalent during rainy weather and under conditions during which the plants are kept wet with irrigation water for extended periods of time. Dense canopies created by spacing plants too closely or by insufficient pruning are especially susceptible to outbreaks of botrytis. The organism lives on dead branches, fruit, and foliage, moving to healthy tissues during wet conditions. Splashing rain and irrigation water spread the disease from infected to adjacent tissues. Infected plants are weakened and can be killed. Main and lateral stems can be girdled. Leaves become discolored, taking on a yellowish-gray color at the margins, and may fall prematurely. Infected berries turn brown and rot. Infected tissues can be covered with a soft gray feltlike mass that may contain minute black spots.

The best control for botrytis is to reduce free moisture on the plants and open them up to provide good air circulation. When using overhead irrigation systems, irrigate early in the morning. Use plant spacing recommendations from Chapter 8 and prune the bushes to create open canopies. Ensure that fields are well drained. Good sanitation practices, including removing prunings and other debris from fields, can help to reduce the incidence of the disease. Excessive nitrogen fertilization creates lush growth that is susceptible to the disease and inhibits air circulation. Older reports indicate that bushes suffering from potassium deficiency are more susceptible to botrytis than healthy plants.

### *Cane Blight or Wilt (*Botryosphaeria ribis *Gros. and Dug.)*

This fungus attacks the wood of *Ribes,* apples, roses, and probably other species. Red currants appear to be the most susceptible *Ribes.* Unless controlled, the disease can become very serious. During the late 1800s and early 1900s, 25 to 50 percent of the bushes in North American plantings were killed by cane blight. The fungus appears to be indigenous to North America.

Symptoms include canes suddenly wilting and dying during the summer, often as the fruit begins to ripen. The wood and pith of diseased canes are blackened. Later in the season, black, wartlike bodies appear in parallel rows along the canes. Prune out and burn infected canes. Apply Bordeaux or lime sulfur sprays to dormant bushes in the late fall and early spring.

### *Coral Spot or Dieback* (Nectria cinnabarina)

This fungus attacks the canes of all currants and gooseberries but is most common on red currants. Old and neglected bushes are most susceptible. Occasional branches may show wilting soon after leaves develop, or symptoms may not appear until the fruit begins to ripen. The bark of affected canes may be covered with pink pustules. Infection occurs through dead snags, branches, or pruning stubs. Prune out affected canes using sharp shears. Make clean cuts as close to the ground as possible, disinfecting the shears with 70 percent ethyl alcohol or a 10 percent solution of commercial bleach between cuts. Remove prunings from the field and burn them. Keep bushes pruned to open shapes to allow good air circulation. Prune annually, as discussed in Chapter 9. Prebloom sprays of Bordeaux mixes, lime sulfur, or fixed copper help reduce the incidence of the disease.

### *Currant Mosaic*

This virus causes mottling of the leaves of red and white currants. There is presently little documentation on the effects of the currant mosaic virus on black currants, gooseberries, and jostaberries. The American nematode has been identified in spreading the disease. Plant only clean stock that has been produced by a certified nursery.

Rogue out and burn infected plants. Have soil tests conducted to identify nematode pests present in the field and take steps to control the American nematode.

## Gooseberry or Cluster Cup Rust (Puccinia ribesii-caricis)

This fungus attacks the leaves, stems, and fruits of gooseberries. Sedges are the alternate host for the organism, and both *Ribes* and sedges must be present for the fungus to survive. The disease is seldom a problem but can become serious during periods of wet weather. Outbreaks usually occur on wild gooseberries or in abandoned gardens. Infected leaves are thickened in areas where the cups later appear. The spots appear brilliant yellow in May and may become red. Later the aecial cups break through the undersides of the leaves. Similar symptoms occur on the stems, petioles, and fruit. The disease seldom infects many berries on the same bush. To control, eliminate sedges in and around *Ribes* plantings. Use good sanitation practices by removing and burning prunings.

## Powdery Mildew

At least three species of powdery mildew attack cultivated *Ribes*. European powdery mildew *Microsphaera grossulariae* has historically caused few serious problems with currants and gooseberries having European origins and is not, apparently, found in North America. *Sphaerotheca mors-uvae* (Schw.) Berk. and *S. macularis* have both been identified in North America, with the former being more common, at least in the Pacific Northwest (Pscheidt and Ocamb, 2001). As discussed earlier, the North American species of powdery mildew create challenges when growing certain gooseberry and currant cultivars.

These fungi overwinter on gooseberry and currant twigs, attacking shoots, leaves, and fruit. The fungi causing powdery mildew appear as white powdery growths on the surfaces of leaves, green shoots, and fruits. Infected plants are often stunted, and severely affected plants can be killed. As the fruit matures, the mildew changes to a dark brown, feltlike coating that renders the berries unmarketable. Affected leaves develop scorch symptoms, become deformed, and dry out. During hot weather, damaged leaves may fall off.

The best mildew-control strategy is to select cultivars that are re-sistant to the disease organisms. Many European gooseberries are highly susceptible to American powdery mildew and can be grown in North America only with difficulty. Exceptions are described in Chapter 7. European black currants are also attacked by mildew, es-pecially when cold, wet springs are followed by hot, dry summers. White currants, red currants, jostaberries, and American gooseberries can be infected by mildew but typically suffer less damage than Euro-pean gooseberries or European black currants, with the damage usually being confined to the leaves and shoots.

The difficulty with selecting resistant currant and gooseberry culti-vars is that different strains of the disease organisms exist in different parts of North America. A cultivar that performs well in one area may not perform as well elsewhere. It is also not unusual for a planting that has remained healthy for years to suddenly be devastated by a strain of the fungus brought into the planting on infected nursery stock. Before investing heavily in any planting stock, evaluate trial plantings of prospective cultivars at your site for several years.

Good nutrition can provide some degree of protection. Keeble and Rawes (1948, p. 249) recommended reducing nitrogen fertilization but increasing potassium applications to develop "hardened foliage and hard, well-ripened growth less liable to attack."

Mildew control sprays may be combined with leaf spot and goose-berry maggot control programs. Traditional chemical fungicides, such as sulfurs and demethylation inhibitors, have not proven alto-gether satisfactory in controlling powdery mildew on *Ribes*. Myclo-butanil, which is manufactured by several pesticide manufacturers and sold under a variety of trade names, is a relatively new fungicide that is registered for mildew control on currants and gooseberries in at least some North American regions. According to researchers at Clemson University (Anonymous, 2000), powdery mildew strains on cucurbits have developed resistance to myclobutanil in laboratory tests. As is true with demethylation-inhibitor fungicides, growers are cautioned to closely follow label directions and to use a variety of fun-gicides and cultural practices to slow the development of fungicide-resistant pathogens.

Several new integrated pest-management techniques for powdery mildew control are available. Horticultural mineral oil sprays formu-lated for summer applications have been effective in reducing mildew in black and red currants (Hummer and Picton, 2001) and are reported to be effective on gooseberries as well. Although research data on *Ribes* crops are very limited, some reports suggest the oils have at least limited eradicant and prevention properties. In addition, a biological, foliar-applied fungicide formulated with a hybrid fungal organism, *Trichoderma harzianum*, strain T-22 (PlantShield, BioWorks, Inc., Geneva, New York) and a bacterial biofungicide, *Bacillus subtillis* QST731 strain (patent pending, Serenade AgraQuest, Inc., Union Gap, Washington) are available. A biological control containing the fungal hyperparasite, *Ampelomyces quisqua* in a mineral-oil-based surfactant also shows promise as a mildew preventative (Pscheidt, 2000).

## Black Currant Reversion Disease

The causal agent of this disease is black currant reversion associated virus (Lemmetty et al., 1997). The disease affects black currants and is spread by the black currant gall mite. Neither the gall mite nor reversion disease have been reported in Canada or the United States (Dale, 1992). Both are common in other countries in which currants are grown.

Although European black currants are the usual host of reversion virus, it has also been found in *Ribes bracteosum* Dougl., *R. rubrum* L. var. *pubescens* Swartz, and *R. carrierii* (*R. glutinosum* Benth. × *R. nigrum* L.) which had been infected with the virus by gall mites. Several other species and hybrids of currants have been infected by grafting to infected rootstocks. Red currants may also be affected by the disease (Jones, 1992).

Symptoms in black currants appear one year after infection and are generally limited to one or a few shoots. By the second year, one-third to one-half of the bush is affected, and the entire plant becomes affected by the third or fourth year. Symptoms include a reduction in the size and number of primary leaves that subtend flowers. Leaves produced during the blossom period are chlorotic. Infected bushes tend to have more but shorter canes than uninfected plants. Infected sepals are not as hairy as normal ones, causing flower buds to be brightly colored and shiny, as opposed to their normal downy, gray appearance. Leaves on infected plants are flatter than normal, have a smaller separation between the basal lobes (sinus), and have decreased numbers of main and marginal serrations. Healthy leaves have five to eight veins running from either side of the midrib to the margins of the central or terminal lobe. Infected leaves have less than five lateral veins on either side of the midrib. It may require two years from the time of infection for symptoms to develop.

Use of clean stock is absolutely critical. Ensure that black currants are free from reversion and black currant gall aphid before purchasing the stock. Entry restrictions and quarantines should be strictly observed. Inspect bushes frequently, especially just before flowering when infected buds can be identified. Rogue out and burn infected plants.

### Septoria Leaf Spot [Mycosphaerella ribis (Fuckel) Kleb.]

This fungus infects the leaves of all currants and gooseberries. The disease is most destructive in the Mississippi Valley but has been reported throughout the contiguous United States, Alaska, and Canada. The disease seldom causes severe crop losses, but extensive defoliation can reduce the vitality of the bushes and expose the berries to sunscald. Small, brown spots appear on infected leaves about June. At this time, the symptoms resemble anthracnose. The spots soon enlarge and the central area becomes light in color with a brown border. Small, black specks are scattered over the spots. The diseased leaves, especially on currants, turn yellow and drop.

Practices to control anthracnose leaf spot are usually effective in controlling Septoria leaf spot. Remove dead leaves or cover them with soil by cultivating before new leaves develop. Prune to open the centers of the bushes and provide good air circulation. Septoria leaf spot control is usually combined with powdery mildew control programs. Chemical controls are available in some areas.

### Viruses

Approximately 14 virus or viruslike diseases have been reported on *Ribes,* the most serious being reversion disease (Jones, 1992). These diseases, along with their causal agents and vectors are listed in Table 10.1.

TABLE 10.1. Virus and viruslike diseases of *Ribes*.

| Disease | Causal agent | Vector |
|---------|-------------|--------|
| Reversion | Black currant reversion associated virus | Black currant gall mite |
| Interveinal white mosaic | Alfalfa mosaic virus | Aphid |
| Vein banding | Possible badnavirus | Aphid |
| Mottle | Cucumber mosaic virus | Aphid |
| Wildfire | Not known | Aphid |
| Leaf pattern | Tobacco rattle virus | Nematode |
| Green/yellow mottle | Arabis mosaic virus | Nematode |
| Mosaic | Tomato ringspot virus | Nematode |
| Spoon leaf | Raspberry ringspot virus | Nematode |
| (Usually latent) | Strawberry latent ringspot virus | Nematode |
| (Usually latent) | Tomato black ring virus | Nematode |
| Yellows | Not known | Not known |
| Infectious variegation | Not known | Not known |
| Full blossom | Phytoplasma | Leafhopper |

*Source:* Jones (1992).

# Chapter 11

# Harvesting, Storing, and Marketing *Ribes* Crops

## *HARVESTING*

Harvesting berries by hand is a labor-intensive process that accounts for 60 to 70 percent of the labor needed for these crops. The berries are harvested either by hand or machine. By hand, a person can harvest about 9, 18, or 22 pounds (4, 8, or 10 kg) per hour of black currants, red currants, or gooseberries, respectively (Harmat et al., 1990).

Due to high labor costs, commercial black currant producers in Europe began developing mechanical harvesters during the 1960s and 1970s. Brennan (1996, p. 191) reported that "[t]he crop is increasingly harvested mechanically, and in the United Kingdom virtually all black currants are harvested by machine." Prospective black currant growers should carefully consider how their crops will be harvested before investing in planting stock. Machine harvestability is an important consideration in modern black currant breeding programs, but many older and some new black currant cultivars are unsuited for mechanical harvesting. With some cultivars, the berries do not separate readily from the canes and harvest yields are low. With other cultivars, poor cane anatomy and strength can lead to excessive injury during mechanical harvesting. Typical injures include broken shoots or rubbing off of the bark (Salamon, 1993). There are several basic harvester designs, not all of which are suitable for all cultivars and plant spacings. Growers will find it advisable to contact machine harvester manufacturers for recommended cultivars and field layouts before planting.

Gooseberries can also be harvested mechanically but present some difficulties. Salamon and Chlebowska (1993) reported on preliminary trials with mechanical harvesting of gooseberries, noting that thornlessness and large berry size are important characteristics for successful machine harvesting. Thorns puncture the fruits during the shaking process, and large fruits appear to separate more easily from the canes than smaller berries. Results were impressive, however, with estimates of 1,100 pounds (500 kg) of gooseberries harvested per man-hour, compared with 13 to 35 pounds (6 to 16 kg) per man-hour with handpicking. As is true with most crops, mechanically harvested gooseberries are usually better suited to processing than fresh use.

Red and white currants can be harvested mechanically, but the soft berries are more easily damaged than black currants or gooseberries and are suited only for processing. Mechanically harvested red and white currants contain many damaged berries that rot quickly and must be frozen or processed as quickly as possible after harvest to avoid serious losses. Besides damage to the fruit, red and white currant canes can be injured during mechanical harvesting. In Finnish trials, Tahvonen (1979) observed no significant damage to black currants but extensive injuries on red currant canes after mechanical harvest. Besides the damage done directly to the canes by shaking, wounds created on the stems became infected with *Botrytis cinerea,* leading to dieback of the canes. *Botrytis* infection was especially severe during a cooler-than-average autumn, as was dieback the following spring. Little or no information has been published on mechanical harvesting of jostaberries. Given a sprawling habit and very large canes, machine harvesting may be difficult for this crop.

Many mechanical harvester designs have been tested, including shakers, vacuums, and air blasters. The simplest machine-assisted harvester consists of a battery-operated, handheld vibrator that pickers press against the canes to shake off ripe fruits. Baskets or tarps underneath the bushes catch the fruits as they fall.

The next level of mechanical harvesting involves small machines that harvest half of a row at a time. These small machines were evaluated in Poland and proved valuable on small black currant farms (Felski and Brzezinska, 1988). For traditionally designed and hand-harvested fields, handpicking accounted for 85 percent of the total labor input. Labor for hand-harvested crops accounted for 50 percent of the total production costs. Use of the half-row harvester reduced man-hours per tonne of harvested fruit from 332 to 30 hours. An over-the-row machine tested at the same time reduced labor still further, to 17 man-hours per tonne. Machine harvesting in these trials reduced labor to between 9 and 18 percent of total production costs. Labor for mechanically harvested fields was primarily associated with pruning, which accounted for 60 percent of the total labor expenditure. Felski and Brzezinska (1988) recommended the use of half-row harvesters for small, family farms for which investment capital was limited, since at the time the half-row harvesters cost one-third as much as larger one-row models.

Most currant and gooseberry harvesters marketed today are either tractor-driven or self-propelled machines that straddle the crop rows and harvest berries continuously as they travel down the rows. Vibrating fingers or rods penetrate into the bushes, shaking the berries free from the canes. The berries then drop into catch pans or plates just above the ground, and conveyor belts carry the fruit to storage containers, usually passing first through a forced-air device to remove debris. Some harvesters also use wagons or trailers drawn alongside the main machine to collect fruit.

Combine-style over-the-row harvesters have also been developed for black currant harvesting in alternate-year systems (Olander, 1993). These machines cut the canes off near the ground as they pass over the row or through a meadow planting. As the canes pass through the machine, the berries are shaken off and collected. In Olander's tests (1993), removal of more than 99 percent of the berries was possible. The system works for black currants, which produce significant crops on second-season canes, but would probably not be economical on gooseberries or red or white currants, which produce their best crops on third- or fourth-season canes.

Many factors are involved in the removal of fruit from bushes with mechanical harvesters. As the canes are shaken, multiple and variable breaks develop in the fruit stems and racemes (Utkov and Pilenko, 1984). Canes, however, are not only flexible, but their flexibility varies according to their length, diameter, and other factors. The length of the shaker fingers, their range of motion and vibrational frequency, the amount of time they remain in contact with the stems, stem flexibility and length, fruit ripeness, and weather conditions are just a few factors that come into play. Speed over the row must be carefully controlled. Moving too swiftly reduces the time the fingers contact an individual cane and can lead to insufficient fruit removal. Moving too slowly can increase damage to the canes and increases labor and machine costs.

Because *Ribes* production is very limited in North America, few commercial currant or gooseberry harvesters are available here. Commercial harvesters are or have been manufactured in the United Kingdom, Finland, Hungary, Denmark, Germany, and Poland. As of 1990, the most popular continuous harvester was manufactured by the Pattenden Co. in the United Kingdom (Harmat et al., 1990). Mechanical currant and gooseberry harvesters can be imported from Europe, or growers may be able to modify North American–built units designed to harvest other small fruits, such as blueberries.

In Europe, currants and gooseberries to be mechanically harvested are sprayed with chemicals that cause the plant tissues to produce and emit ethylene. Ethylene is a naturally occurring chemical produced in plants and has hormone-like properties, including speeding up ripening of fruit. Another important property of ethylene is that it causes an abscission layer to form between a fruit and its stem in some crops. This layer quickly dries out and breaks, allowing the fruit to be easily removed from the plant.

Commercially available products are commonly applied to apples and other fruit crops to shorten the time to ripening and ripen most or all of the fruits at once. Uniform ripening is very important in mechanical harvesting currants and gooseberries for which it is desirable to pick all of the fruit at one time but for which flower fertilization and fruit development can occur over as much as 20 days. In Hungarian trials, Zatyko and Sagi (1972) tested a commercial ethylene-producing compound, Ethrel [also known as ethephon or (2-chloroethyl) phosphonic acid], on black and red currants. In these trials, Ethrel applied to black currants at 480 ppm allowed more than 98 percent of the berries to be shaken from 'Boskoop Giant' and 'Blacksmith' canes four days after application, compared with less than 7 percent of the berries on untreated plants. Results were similar to those of red currants, producing fruit drops of 93 percent and more than 98 percent in 'Hosszufurtu Piros' and 'Jonkheer van Tets', respectively, compared with 12 and 56 percent fruit drops on untreated plants. Whereas the black currants dropped without strigs, approximately 10 percent of the red currants remained attached to the strigs. Costin and Kenny (1972) tested Ethrel on gooseberries in Ireland and concluded that up to 480 ppm of Ethrel could be applied to 'Careless' without causing deterioration of fruit quality, but 360 ppm Ethrel produced the least overall effects on fruit quality and 240 ppm would probably be the most suitable concentration for commercial applications. A standard practice in Europe now for mechanically harvested crops is to apply 240 ppm of ethephon to black and red currants about ten to twelve days before harvest to increase harvest speed and uniformity (Harmat et al., 1990).

Although ethylene-inducing chemicals can be helpful in harvesting, they also represent risks to crops. Ethylene is sometimes referred to as a wound hormone and can damage or kill plant tissues when used improperly. Follow label rates and instructions carefully to reduce the risk of crop damage. Be prepared to harvest when you apply the chemical, and treat no more acreage than you can harvest in a two- or three-day period. Once the abscission layers form, fruits will rapidly begin to drop.

Ethephon products also create risks for humans and other animals. Although the toxicity is relatively low, ethephon is corrosive and can cause irreversible eye damage. Ethephon carries a "Danger" classification and is harmful if swallowed or absorbed through the skin.

Ethephon is used in Europe for harvesting currants and gooseberries, but it may not be available for growers in North America. As of December 2002, the Ethrel brand of ethephon was labeled in the United States for tomatoes, cherries, grapes, apples, walnuts, peppers, blackberries, blueberries, and cantaloupes. No mention was made regarding legal use on currants or gooseberries. Before applying ethephon or any other growth regulator, growers must ensure the products are registered for their crops and locations. Manufacturers and farm chemical suppliers can provide labels and registration information on ethephon products. If a product cannot be found that is labeled for currants and gooseberries, it may be possible to obtain a waiver or special application permit through your state or provincial department of agriculture.

### Harvesting Currants

Currants ripen over about a two- to four-week period, depending on the cultivar and weather. They can be harvested in two pickings to get the berries at peak ripeness, although most currants and gooseberries remain on the bushes without falling or becoming overripe for one to four weeks after ripening (Van Meter, 1928; Harmat et al., 1990). Berries ripen as early as May in warmer locations, and some cultivars are picked in August in cooler climates. Cold weather during spring and early summer can delay ripening.

For home use and local, fresh-market sales, allow currants to ripen fully before picking. When fresh berries must be firmer for shipping, harvest when the fruits are mature but have not fully ripened and become soft. Currants destined for jellies and other preserves are often picked slightly underripe because the fruit pectin content is highest at that time. In North America, however, the tendency has been to pick currants too early because of the misconception that ripe fruits do not jell properly during cooking. Ripe berries are sweeter and taste better than green ones and can be used effectively for processing. When ripe, the currants will be soft, flavorful, and fully colored with no trace of green on the stem ends. Overripe berries shrivel and become mushy. Since fruit for juices and jellies is strained before processing, the stems do not have to be removed before processing (Harmat et al., 1990).

Red currants are generally harvested before the skins change from bright to dull red (Audette and Lareau, 1996; Spayd et al., 1990), and white currants should also be harvested while the skins remain bright and transluscent. Total soluble solids (TSS) will normally fall between 9.5 and 14 percent when red and white currants are ready for harvest (Prange, 2002). Red and white currants have delicate skins and, when hand harvested, are collected in intact clusters by carefully pinching the strigs off at their bases. If the berries are stripped from the clusters, the skins tear and the berries quickly spoil. Pickers must take care not to crush the top berries when pinching off strigs. The juice from crushed berries supports fruit rot organisms, such as botrytis, and a few crushed and rotting berries can infect all of the fruit in a container. To prevent fruit loss, damaged berries should not be kept, even for processing. Cultivars with long stems are easier to pick without damaging the berries, which is why they are generally preferred for fresh-market sales.

Black currants should be uniformly black or very dark blue with no trace of green on the skins at the time of harvest. Soluble solids can range from 15 to 26 percent in ripe black currants (Prange, 2002), and growers will need to gain experience with their crops to determine the optimum harvest times for their particular cultivars and markets. Because black currant fruits on an individual strig do not ripen at the same time, the berries may be picked by hand over two or three pickings to remove individual, ripe berries. With machine harvesting, all of the berries are picked at one time. Black currants are firmer than red or white currants and can be picked individually or in clusters.

Fruit rot problems greatly increase when harvested fruit is wet. Pick and pack fruit only when it is dry. Never pick in the rain or wash berries prior to or during packing. As soon as a flat or container of berries is full, it should be set in the shade to keep the fruit cool. Full flats should be collected frequently and transported to a cold room or refrigerator.

For fresh use, pick currants into one-half or one-pint (250 to 500 mL) containers. Leave red and white currant clusters intact. Avoid mesh baskets for currants. The berries catch in the mesh, tear, and leak. Solid baskets and clear, plastic, clamshell containers designed for raspberries, strawberries, and blueberries work well for fresh currants. If the fruit is destined for processing, handpick into one- to four-quart (1 to 4 L) containers. The collection containers used for mechanical harvesters depend on the type of harvester and the intended use of the fruit. Currants weigh about 1.2 pounds per quart (575 g/L).

## Harvesting Gooseberries and Jostaberries

Gooseberries begin ripening at about the same time to several weeks later than currants, depending on cultivar, and are harvested over a four- to six-week period. Colors of ripe fruits vary dramatically among cultivars and include green, white, yellow, pink, red, purple, blue, and nearly black. Because there are markets for immature and fully ripe berries, harvest maturity depends on the intended use (Ryall and Pentzer, 1982). In England, processing berries are often harvested when they reach their maximum size but are still slightly green and quite tart (Harmat et al., 1990). This practice is referred to as the "green berry trade." For better flavor, processing gooseberries can be allowed to ripen or nearly ripen before harvesting. When they are hand harvested, gooseberries are generally picked about three times each season. During the first picking, harvest one-third of the fruit, leaving the remainder evenly distributed on all the canes. Thinning the fruit increases the size of the remaining berries by reducing the demand for food reserves. During the second picking, strip berries from low-lying canes and those canes in the center of the bush, leaving the best-quality fruit to ripen on well-exposed outside branches (Keeble and Rawes, 1948). Berries from the first two pickings are normally used for processing, while those from the third picking are suited for fresh use.

Gooseberries are firm and resist bruising when green but soften as they ripen. Be more careful when picking ripe than underripe fruit. Because of their size, firmness, and the presence of thorns on many cultivars, gooseberry harvesting tends to be rather rough, resulting in leaves and other debris being collected into the picking containers. Gooseberries for processing can be harvested by stripping them from the bushes with heavy gloves or cranberry scoops. This method is not well suited for fresh-market fruit because the berry skins are torn and punctured by thorns (Shoemaker, 1948). Fresh-market berries must be picked more carefully. Some pickers wear long, leather or canvas aprons that are stretched out under the bushes; they collect the fruit into their aprons before transferring it to containers. Gooseberries are occasionally harvested by placing tarps under the bushes and "cuffing" the berries off by striking the canes with wooden bats or paddles.

Berries for fresh or dessert use are picked individually by hand into one-pint or one-quart (0.5 to 1 L) containers. Processing berries can be picked into one- to four-quart (1 to 4 L) baskets or flats. To remove debris, pour the berries gently from one container to another in front of a fan or run the fruits over commercially available cleaning tables that use screens and forced air to clean the fruit. Jostaberries are harvested the same way as gooseberries. As with currants, do not harvest wet fruit, keep full containers in the shade after picking, and transport the fruit to a cold room or refrigerator as soon as possible after picking. Gooseberries and jostaberries weigh about 1.25 pounds per quart (600 g/L).

### Grade, Size, and Packaging Standards

No U.S. standards have been developed for fresh currants, gooseberries, or jostaberries, although there is a U.S. grade standard for processing currants based on color, stem attachment, freedom from decay, and insect or mechanical damage (Prange, 2002). Processors tend to set their own standards for these fruits. For fresh markets, shipping and display containers are generally the same as are used for blueberries or similar fruits.

## POSTHARVEST CARE AND STORAGE

### Refrigeration

Currants, gooseberries, and jostaberries should be cooled as quickly as possible after picking, especially when intended for the fresh market. Hydrocooling (immersing fruit in cold water) is commonly used for tree fruits but increases berry rot problems and should not be used for currants, gooseberries, or jostaberries. Instead, forced air at a relative humidly of 95 percent is used to cool the fruits (Batzer and Helm, 1999; Kasmire and Thompson, 1992). The accepted method of cooling *Ribes,* as well as raspberries and other small fruits, is to harvest them into small containers that sit in shallow flats designed to allow air to circulate around the fruit. The flats are stacked in front of fans in walk-in coolers. Tarps are laid over the tops and ends of the stacks of flats to force cold air through vent holes in the flats and through the fruit.

Currants, gooseberries, and jostaberries are not chilling sensitive (Kader, 1992) and should be cooled to and held at 31.1 to 32°F (–0.5 to 0°C) at 95 percent relative humidity (Hardenburg et al., 1986; Story and Simons, 1989). Batzer and Helm (1999) recommended somewhat warmer storage temperatures for currants and gooseberries, perhaps to avoid accidental freezing. Their recommendations were 32 to 34°F (0 to 1°C) for red currants and gooseberries and 32 to 36°F (0 to 2°C) for black currants. An important practice to maintain high fruit quality is to cool the berries to about 34°F (1°C) within two to four hours of harvesting. With proper cooling, black currants can be stored for 1.5 weeks, red and white currants for 2.5 weeks, and gooseberries for three weeks using conventional refrigeration (Batzer and Helm, 1999). Refrigeration is discussed in Chapter 5.

### Controlled-Atmosphere Storage

According to Batzer and Helm (1999) and Thompson (1998), gooseberries and red and white currants respond very well to storage under controlled-atmosphere (CA) conditions, although black currants benefit only slightly from CA storage. CA storage has been used commercially for apples and other fruits for many years. The advantage of using controlled atmospheres is that storage times can be greatly increased without loss of fruit quality. With currants and gooseberries, modifying the carbon dioxide ($CO_2$) and oxygen ($O_2$) concentrations extends storage life.

The earth's atmosphere at sea level normally contains about 20.95 percent oxygen and 0.03 percent carbon dioxide, with nitrogen being the most abundant gas at 78.08 percent. Argon makes up about 0.93 percent of the atmosphere, with trace amounts of neon, helium, krypton, hydrogen, xenon, ozone, methane, water vapor, and several other gases also present (Morgan et al., 1980). Carbon dioxide and water vapor concentrations can vary significantly in different locations and over time. Increasing the $CO_2$ concentration up to 20 percent helps reduce the incidence of storage rots (Batzer and Helm, 1999; Thompson, 1998), while lowering the $O_2$ concentration slows respiration of the harvested fruit (Robinson et al., 1975). Great care must be taken to monitor and control conditions in CA chambers because too much or too little $CO_2$ or $O_2$ can damage the fruit or shorten storage life.

Humidity levels are kept high (95 percent RH) in refrigerated and CA storage to reduce losses of fruit yields and quality through desiccation. Regardless of storage system, some water loss from fruit is inevitable. Roelofs and Waart (1993) found in red currants that drying through the fruit stems and other water losses were high during the first eight to eleven weeks of storage. Although high humidity is important, moisture on the fruit skins is undesirable because it encourages activity by rot organisms. Air inside a refrigerated or CA storage unit should circulate constantly to prevent accumulation of free moisture. Keeping storage temperatures slightly above freezing slows fruit ripening and retards fungal and bacterial pathogens. Care must be taken in monitoring and controlling storage temperatures. Allowing currants, gooseberries, or jostaberries to freeze destroys their fresh-market value.

In some CA storage systems, control of ethylene concentrations is very important. As discussed earlier in this chapter, ethylene speeds the ripening of many fruits. Furthermore, some fruits generate ethylene as they ripen. For example, as apples ripen, they generate ethylene that causes them to ripen even faster. For that reason, chemicals and apparatus are used in apple and some other fruit storage rooms to scrub ethylene from the atmosphere and slow fruit ripening while in storage. Unfortunately, literature searches by the authors and others have failed to uncover any data on ethylene production by, or its effects on, currant, gooseberry, or jostaberry fruits.

Red and white currants can be stored for 8 to 14 weeks (depending on cultivar) at 33.8°F (1°C), 18 to 20 percent $CO_2$, and 2 percent $O_2$ (Prange, 2002). Roelofs and Waart (1993) found that fresh fruits from five red currant cultivars retained "reasonable quality" for more than 20 weeks at 33.8°F (1°C), 20 percent $CO_2$, and 2 percent $O_2$. This compares with 2.5 weeks using refrigeration alone. According to Prange (2002), gooseberry storage can be extended to six to eight weeks in CA storage at 33.8°F (1°C), 10 to 15 percent $CO_2$, and 1.5 percent $O_2$. Acceptable quality for fresh black currants can be maintained for up to three weeks by storing the berries at 32 to 35.6°F (0 to 2°C) with 15 to 20 percent $CO_2$.

Excessively high $CO_2$ concentrations can damage fruit. Roelofs and Waart (1993) found that some red currant cultivars suffered from discoloration and internal breakdown after 13 weeks in storage at $CO_2$ concentrations above 20 percent. Low oxygen levels further increased these symptoms. Green gooseberry fruits can become yellow and develop abnormal flavors when held in air at 32°F (0°C) when $CO_2$ concentrations were increased from 8 to 12 percent (Smith, 1967). Increasing the temperature to 41°F (5°C) eliminated the disorder.

Time of harvest is an important factor in maintaining fruit quality during storage. The quality of overripe fruits does not improve regardless of the storage system used. It is better to pick berries for refrigerated or CA storage slightly underripe rather than overripe. Roelofs and Waart (1993) also found with red currants that the earliest-ripening berries were more sensitive to rot than later-ripening fruits, regardless of cultivar. Growers and brokers may wish to market red and white currants from the first pickings quickly, reserving long-term storage for berries from later harvests.

Although $CO_2$ generators and scrubbers are commercially available, they are not always necessary for storage of berries. In recent years, fruit-packing houses have begun using special, semipermeable plastic films for CA storage of blueberries. Flats of blueberries are stacked together in units approximately 4 × 4 × 8 feet high (1.3 × 1.3 × 2.6 m) and the units wrapped with the semipermeable film. The wrapped units are then secured to standard, wooden pallets for easy transport and the palletized flats are stored in a refrigerated room. Inside the plastic film, the berries continue to respire, increasing the $CO_2$ concentrations and decreasing the $O_2$ concentrations to desired CA levels. It is possible that the same technique can be applied to currants, gooseberries, and jostaberries.

Many commercial controlled-atmosphere systems are available. Before selecting a system and investing in facilities and equipment, growers and brokers should visit several commercial fruit-packing houses and contact several equipment manufacturers to determine which storage approaches are best for their particular operations.

### Postharvest Problems

The main postharvest disease in *Ribes* fruits is gray mold *(Botrytis cinerea)* (Prange, 2002), which can appear as small, brown spots on the berries (Dennis, 1983; Harmat et al., 1990; Ryall and Pentzer, 1982). These spots enlarge rapidly at temperatures above 50°F (10°C), causing soft rot on infected fruits (Prange, 2002). Berries borne near the ground can be contaminated by splashing rain or irrigation water with the fungal pathogen *Mucor piriformis* (Dennis, 1983). In gooseberry, *Alternaria* and *Stemphyllum* fungi can infect the seeds contained within the pericarp of the fruit (Dennis, 1983). When infected fruits are stored at room temperatures for several days, the pathogens can invade the pericarp and rot the berries.

High quality in harvested fruit begins with careful management during the growing season and skilled pickers. Berries infested with currant maggot, gooseberry fruitworm, or currant moth larvae are subject to rot and can infect other berries during transport or storage. Maggot-infested berries are also not particularly palatable to consumers. Berries damaged by bird feeding are also highly susceptible to rot organisms and decrease the value of fresh-market berries, in particular. Likewise, mildewed berries do not store well and reduce the value of fresh and processing fruits.

## MARKETING

Perhaps the greatest challenge *Ribes* growers in North America face is marketing. Aside from a few serious pests and diseases that can be controlled through good cultivar selection and management, the crops are adaptable to a wide range of sites and fairly easy to grow. The amount of labor needed to grow currants, gooseberries, and jostaberries is about the same as for blueberries and less than for many other small fruit crops. Since *Ribes* fruits hang well on the bushes without overripening, harvest windows are longer and provide more flexibility than strawberries, raspberries, or blackberries. Currant, gooseberry, and jostaberry fruits are flavorful and suitable for a range of uses. Red and white currants are visually attractive, and black currants are rich in health-related compounds.

The main challenge in marketing currants, gooseberries, and jostaberries is that relatively few people in North America know what they are with any accuracy or have ever tasted them. Although *Ribes* fruits were popular in North America at one time, lengthy restrictions on their production caused them to be lost from our culture. Even worse, many people that have heard of currants and gooseberries think of them as "those things that kill pines." The belief is not strictly true, but perception is often more important than reality in marketing.

Another challenge *Ribes* growers face is stiff competition. During the golden age of currant and gooseberry production in North America, growers faced little commercial competition in the fruit market. Raspberries and strawberries were common, but blueberries were just being domesticated and were available only in small quantities. Deciduous tree fruits were commonly available throughout Canada and the United States, but transportation technology at the time limited the market for citrus and other tropical and semitropical fruits. Today, any of these commodities, along with kiwi, guavas, papayas, star fruit, and other exotic tropical fruits can be found virtually everywhere.

Consumer education plays a key role in marketing *Ribes*. Growers must convince prospective consumers that the fruits are nutritious and delicious as well as convincing them that the crops do not threaten native and ornamental pines.

Perhaps the strongest marketing advantage *Ribes* growers have lies in the health benefits of consuming currants, gooseberries, and jostaberries. This advantage particularly applies to black currants, which contain high concentrations of anthocyanins and other antioxidants, as discussed in Chapter 2. Given an aging population with demonstrated interest in natural, nutraceutical products, *Ribes* producers and marketers in North America have unique marketing opportunities for fresh fruit and both culinary and dietary supplement products. Local niche markets featuring fresh fruits and value-added products offer good opportunities for small-scale producers.

Developing an industry marketing association, such as was done for low-bush blueberries (also known as wild blueberries), appears highly advantageous. Marketing associations, such as the Wild Blueberry Association of North America, collect research and marketing information, develop promotional materials, issue news releases, maintain informational Internet sites, and otherwise assist individual growers and sellers with marketing. The International Ribes Association (TIRA) was formed in about 1989 and remained in operation at the time of this writing. TIRA provides educational and technical support for *Ribes* growers and processors and regularly sponsors technical conferences in North America.'

## The Product

Recent research has demonstrated the importance of naturally occurring compounds in maintaining and improving human health. Increasing epidemiological evidence associates diets rich in fruits and vegetables with reduced risk of heart disease, cancer, and other chronic diseases. One major benefit from a diet containing fruits and vegetables may be the increased consumption of antioxidants, including such compounds as carotenoids, vitamin C, vitamin E, and phenolics. These compounds can reduce the amount of oxygen radicals, which are known to cause disease.

Dark-pigmented fruits are particularly rich in antioxidants. As discussed in Chapter 2, black currants have very high levels of total phenolics, total anthocyanins, and high antioxidant capacities in comparison with other fruits and vegetables. Some black currant cultivars have anthocyanin levels between 350 and 450 mg per 100 g berries, equivalent to that of bilberry and other blueberry fruits, but not as high as the more than 600 mg/100 g berries of black raspberries (Moyer, Hummer, Wrolstad, and Finn, 2002; Moyer, Hummer, Finn, et al., 2002). Some black currant cultivars contain about four times as much vitamin C as do citrus fruits (Westwood, 1978). Because of the high levels of vitamin C and other compounds, the total phenolics in *Ribes* can reach 1700 mg/100 g berries, a higher level than that of black raspberries or blueberries, both about 1000 mg/100 g (Moyer, Hummer, Wrolstad, and Finn, 2002; Moyer, Hummer, Finn, et al., 2002). *Ribes* fruits also contain pectins, various mineral elements, and fructose. Black currant seeds are rich in essential fatty acids, such as gamma-linolenic acid (Brennan, 1990).

Currants are mostly used in North America for making jellies, relishes, and juices. Red and white currants can be crushed and mixed with water and sugar to produce a lemonade-like drink called currant shrub. Because they are tart, red and white currants can be mixed with less acidic fruits when making preserves. One drawback to white currants is their lack of color in preserves. According to Card (1907, p. 351), "[I]f white currants are scalded before pressing, they make a rich, red jelly, not as dark as that from red varieties, but very handsome." Red and white currant juices were also once thought to be useful in soothing fevers. Although their large, numerous seeds limit fresh consumption, the tart flavor of fresh, ripe red and white currants is very good. White currants are also good when eaten fresh but are generally considered less flavorful than red currants and are used primarily in Europe for baby food (Harmat et al., 1990). White currants are also used in Finland for wines and sparkling wines. Red and white currants make attractive additions to garnishes and culinary art.

The leaves and buds of black currants are rich in phenolic compounds and have long been used for herbal medicines. Card (1907) reported that black currant products were especially useful in treating sore, inflamed throats and were also said to be useful in treating "bowel and summer complaints." In Russia, the leaves have been used in making an herbal tea that resembles green tea in flavor. Black currant juice can be used alone, mixed with other fruit juices, or used to make wines and liquors. Because of their strong, "foxy" flavor, fresh black currants have never been widely popular in the United States but do have some following in Canada. As discussed in Chapter 7, some black currant cultivars are quite palatable when eaten fresh. One way to increase the appeal of black currants is to blanch them in boiling water for several minutes to remove some of the flavor and dark pigment. Rinse the berries and cook in fresh water to lighten the color and further subdue the flavor. Black currant juices and wines are very popular in Europe and are sold occasionally in North America, usually mixed with the juice of other fruits.

Gooseberries are used primarily in jams, jellies, pastries, and compotes. Bush-ripened berries eaten out of hand or as a table dessert are delightful treats. Ripe gooseberries have a sweet, delicate taste and a texture resembling grapes. A few gooseberry cultivars produce berries covered with fuzzy hairs that give fresh fruit an unpalatable texture. Most popular cultivars, however, have smooth fruits. During the 1800s, 'Houghton' gooseberries were used as a substitute for cranberries (Card, 1907). The berries were cooked enough to cause the skins to burst and were then canned unsweetened. The relish was sweetened just before use.

Jostaberry fruits are smooth, sweet, mild flavored, soft, and dark red to nearly black. They are normally used for processing but can be eaten fresh. Like currants and gooseberries, jostaberries keep well when frozen.

As mentioned earlier, consumer education is critically important for the development and expansion of a North American *Ribes* industry. Given currants' and gooseberries' long absence from the market, most consumers have little experience in the preparation and use of these fruits. Providing customers with recipes, serving suggestions, and other information can help boost sales. Newspaper and magazine articles providing recipes for currants, gooseberries, and jostaberries should also help boost demand for these fruits. One source of information is the CD-based collection of recipes *Still in Our Memories: Currants and Gooseberries* (Rudowski, 2002), available from The International Ribes Association and containing more than 800 recipes. Cookbooks from the late 1800s and early 1900s also often contain recipes for currants and gooseberries.

# Chapter 12

# Breeding Currants, Gooseberries, and Jostaberries

Currant and gooseberry breeding date back for centuries in both Europe and North America. Although some limited work is being done today in Canada, little or no university-based currant or gooseberry breeding is being conducted in the United States. The vast majority of *Ribes* breeding programs are based in Europe, where these crops are widely grown. With a long history of cultivation, the availability of many *Ribes* species not yet utilized in breeding, experience with pests and diseases, and rapidly increasing knowledge of *Ribes* cytogenetics, the prospects for improved currant and gooseberry cultivars are promising. The main goals in leading *Ribes* breeding programs are to

- develop disease resistance, especially to powdery mildew, blister rust, reversion, and leaf spot;
- develop resistance to black currant gall mite (a vector of reversion disease);
- develop resistance to late spring frosts, particularly for black currants;
- improve fruit and juice quality;
- decrease variability in yields of black currants; and
- improve mechanical harvesting characteristics.

Brennan (1990, 1996) and Keep (1975) from the United Kingdom have discussed breeding currants and gooseberries. Some of their suggestions, as well as those of the authors, are discussed here. For a detailed treatment of *Ribes* genetics and breeding, readers are referred to "Currants and Gooseberries" (Keep, 1975), "Currants and Gooseberries *(Ribes)*," (Brennan, 1990), and "Currants and Gooseberries" (Brennan, 1996). Suggested cultivar and species donors that may be useful in breeding for specific traits are listed in Table 12.2 at the end of this chapter.

## SELECTION OF PARENTS FOR BREEDING

From the perspective of conventional fruit breeders, selecting parents based on their observable characteristics generally gives good results, as *Ribes* offspring generally closely resemble their parents. Wilson and Adam (1966) concluded that most of the differences observed in offspring could have been predicted from the parents' phenotypes (observable characteristics). In other words, at the beginning of a breeding program, select parents that show the strongest expression of desired characteristics. During evaluation of progeny, continue to select individuals that demonstrate desirable characteristics most strongly.

### Gooseberries

For centuries, gooseberry breeding efforts were primarily devoted to increasing fruit size. Beginning with an average fruit weight of about 0.3 ounces (7.1 g) in the wild species, by 1852 gooseberry enthusiasts in England had achieved a record-setting berry weight of 1.9 ounces (53.9 g). This works out to an increase in fruit size of nearly 800 percent and a berry almost two inches (5 cm) in diameter. Unfortunately, similar efforts were not directed toward developing pest and disease resistance.

Although dwarfed today by most black currant breeding programs, some gooseberry breeding continues. During the 1900s, efforts were primarily directed toward combining the large fruit size and quality of European cultivars with the powdery mildew resistance of American gooseberries (Brennan, 1996). Bauer (1955) utilized the North American species *R. divaricatum* in repeated backcrosses with large-fruited European gooseberries to develop 'Resistenta', 'Perle von Muncheberg', and 'Robustenta' during the 1950s. Later German introductions include 'Remarka', 'Reverta', 'Risulfa', 'Ristula', and 'Rokula' (Brennan, 1996). New German gooseberry cultivars include 'Rixanta', 'Reflamba', and 'Rolanda'. Gooseberry breeding in the United Kingdom lead to the release of 'Invicta' [(Resistenta × Whinham's Industry) × Keepsake] and 'Greenfinch' [Careless × (Whinham's Industry × Resistenta)] (Brennan, 1996). Other species used in European efforts to improve disease resistance include *R. leptanthum* and *R. watsonianum*, with *R. missouriense*, *R. oxyacanthoides*, and *R. sanguineum* providing genes for spinelessness. North American cultivars include 'Captivator', 'Downing', 'Glenndale', 'Houghton', 'Jahn's Prairie', 'Oregon', 'Pixwell', 'Poorman', 'Sabine', 'Sebastian', 'Shefford', 'Spinefree', 'Stanbridge', 'Sutton', and 'Welcome'.

For North American culture, the most important goal in gooseberry breeding should be to develop resistance to powdery mildew. Increasing fruit size and decreasing spininess are also important goals. One of the most promising American gooseberry cultivars is 'Captivator', which is highly resistant to powdery mildew and has good fruit and horticultural characteristics. 'Houghton', one of the first American cultivars developed, is also quite resistant to the disease, as are several European cultivars. Suggested donors for powdery mildew resistance are listed in Table 12.2.

'Captivator', 'Houghton', 'Poorman', 'Spinefree', and 'Welcome' may be valuable in decreasing spininess. 'Spinefree' [(F$_2$ of *R. oxyacanthoides* × Victoria) × Mabel] is noted for being thornless and served as the female parent for 'Captivator'. Other suggested donors for spinelessness in gooseberries are listed in Table 12.2.

Improvements in fruit size are likely to come through hybridization with large-fruited European cultivars. Flavor is also an important breeding consideration. Suggested donors for increased fruit size and improved flavor are listed in Table 12.2.

### Black Currants

In North America, resistance to blister rust is, perhaps, the most important short-term goal in black currant breeding, followed by resistance to powdery mildew, improved yield and cropping uniformity, and improved flavor and juice quality. In the long run, developing resistance to reversion disease and black currant gall mite may become important. Although this disease and pest have not yet been found in North America, the probability is that they will inadvertently be imported eventually.

The Canadian cultivars 'Consort', 'Coronet', and 'Crusader' are resistant to blister rust, as is the Russian cultivar 'Primorskij Cempion' and the promising Swedish cultivar 'Titania'. These cultivars will probably be included in the core germplasm for most breeding efforts on this continent.

Increased resistance to American powdery mildew is a high priority in breeding programs in Europe and North America. Breeding for resistance is complicated by the existence of many closely related races of the pathogen and their tendency to mutate, forming new races. As a consequence, cultivars once considered highly resistant to mildew are now susceptible. In black currants, for example, the Finnish cultivar 'Brodtorp' was highly resistant and used as a donor of mildew resistance in the United Kingdom beginning in 1963. Keep (1977) reported that resistance in 'Brodtorp' was only partial, being conferred by a gene designated $M$, and that the resistance was breaking down. Furthermore, mildew-resistant offspring from 'Brodtorp' tended to have a sprawling habit. The Finnish cultivar 'Lepaan Musta' also carries the $M$ gene. The Swedish cultivar 'Ojebyn' is heterozygous for the dominant, mildew-resistant gene $Sph_2$. Although reported by Keep (1977) to be effectively resistant to mildew, Trajkovski and Paasuke (1976) suggest that, as with 'Brodtorp', 'Ojebyn's' resistance may be short-lived. Swedish cultivars 'Sunderbyn II' and 'Matkakowski' carry a dominant, mildew-resistant gene $R$ (Trajkovski and Paasuke, 1976). The newer Scottish cultivars 'Ben Alder' and 'Ben Sarek' also exhibit resistance to powdery mildew. The North American species $R.$ $glutinosum$ and $R.$ $sanguineum$ have high levels of mildew resistance. Keep (1977) reported that offspring of $R.$ $glutinosum$ in the third backcross from that species were resistant to mildew and exhibited high yields and good plant habits. Keep (1985) later reported, however, that the mildew-resistance gene, $Sph_2$, might be repulsion linked with black currant gall mite resistance. In other words, plants resistant to gall mite are less likely than non-mite-resistant plants to carry the $Sph_2$ gene. Keep (1986) also found evidence linking the $Sph_2$ gene to cytoplasmic male sterility. Keep's observations underscore the need for careful screening of progeny. In breeding for mildew resistance, it seems advisable to incorporate multiple genes for resistance, rather than depending upon a single gene.

Few black currant cultivars now commonly grown in North America meet European standards for juice quality. If commercial North American black currant production is to compete globally, juice quality must be improved. Of particular importance are anthocyanins, which give black currants their color and also serve important antioxidant functions in the human body. Biochemical studies over the past 40 years have provided detailed information on the composition of black currants, allowing breeders to select precisely for desirable traits. Chapter 2 describes fruit composition in some detail.

Western European black currant cultivars primarily contain cyanidin derivatives, while those from Scandinavia are delphinidin derivatives (Taylor, 1989). Nordic cultivars 'Brodtorp' and 'Janslunda' served as parents in the development of 'Ben Alder', which contains high concentrations of anthocyanins, making it desirable for processing. The Scandinavian cultivars, apparently, also donated low concentrations (about 3 percent of total) of acylated anthocyanins that retain their color over a wide pH range and are less prone to degradation than other anthocyanins during processing (Brennan, 1996). Another black currant breeding goal might be to increase the proportion of acylated pigments in cultivars, although further studies of *Eucoreosma* species are necessary to identify prospective donors.

'Baldwin' has excellent juice quality and once dominated production the United Kingdom. Frost susceptibility, low yields, and susceptibility to powder mildew make this cultivar a poor choice today, although it is still considered a standard for flavor. Knight (1983) observed that in using *R. nigrum, R. dikuscha, R. bracteosum,* and *R. grossularia* to introduce improved agronomic characteristics, seedlings selected after three to four backcrosses to a range of *R. nigrum* cultivars produced juice of commercial quality. Scottish cultivars 'Ben Lomond', 'Ben Alder', and their offspring produce fruits having excellent juice quality, possibly having inherited the trait from the Scandinavian cultivar 'Janslunda' (Brennan, 1990). Other potential donors possessing commercially acceptable juice include 'Black Reward', 'Greens Black', 'Malvern Cross', 'Roodknop', 'Seabrook's Black', 'Wellington XXX', and 'Westwick Choice' (Knight, 1983).

Flavor is another important consideration in breeding. Anderson (1977) found that the flavor characteristics of 'Baldwin' were seldom found in cultivars from Scandinavia and the former USSR or their derivatives. Hybrids of *R. bracteosum* and *R. ussuriense* resembled even less typical black currants in flavor (Brennan, 1996). Hybrids of *R. petiolare* tend to develop unpalatable flavors (Melekhina et al.,

1980), and backcrossing to *R. nigrum* is required to reduce the undesirable characteristics in the juice.

### Red and White Currants

In terms of breeding, red and white currants need, perhaps, the least improvement of all the *Ribes* crops. The commercial popularity of these crops has waned over the years, and breeding efforts have done likewise. 'Redstart' [Red Lake × (*R. multiflorum* × *R. sativum*)], developed by Elizabeth Keep at the East Malling Research Station in 1990, was the first major red currant release in the United Kingdom since 1910 (Brennan, 1996). Between about 1920 and 1941, 'Red Lake' was released in the United States and 'Jonkheer van Tets' and 'Rondom' in Holland and the Netherlands. Norwegian breeder J. Oydvin (1978) released 'Fortun', 'Jontun', and 'Nortun' in 1978.

The most important goals for red currant breeders might include resistance to powdery mildew and cane blight, higher yields, late flowering for improved frost resistance, and longer handles on strigs to facilitate easy handpicking. White currants are similar to red currants in most respects, although they have the largest and most consistent yields of all cultivated *Ribes*. Harmat and colleagues (1990) reported white currant yields as high as nine tons per acre (20 tonnes/ha). Breeding objectives are the same for white currants as for the red types. Kronenberg (1964) found that flavor varied among red currant cultivars, with desirable flavor relating to high citric acid concentrations compared to tartaric and malic acid concentrations. 'Fay's Prolific' and 'Laxton's No. 1' were noted as having particularly desirable flavors, while the flavors of 'Jonkheer van Tets' and 'Rondom' were considered weaker.

## RIBES *BREEDING TECHNIQUES*

### Conventional Cross-Pollination

As mentioned previously, currants and gooseberries are relatively easy to breed. Although genetic engineering and DNA profiling methodologies are changing the face of *Ribes* breeding, much progress remains possible through conventional cross-pollination. Selection of appropriate maternal parents and pollen donors is of first importance. Little success comes from crossing inferior cultivars. Focus is also important. Breeders should identify one or a few specific traits of interest, rather than trying to select for many traits at once.

Cross-pollination can be carried out inside greenhouses, screen-houses, or in field plots. Greenhouses and screenhouses have the advantages of providing dry working conditions (a rarity in early spring for many growers), protection from spring frosts, and excluding bees and other pollinators that could interfere with controlled crosses. Field crosses are entirely feasible, however, as long as breeders take care to prevent unwanted pollinations by bagging flower clusters to keep out insects. Do not apply sulfur or thiodan fungicides to open flowers, as such applications have been found to greatly reduce pollen viability in black currants (Keep, 1975).

Currants and gooseberries typically flower over a three- to four-week period beginning in April or May. Flowering, of course, occurs earlier in warmer climates. Individual flowers remain open and the stigmas receptive to pollen for five to six days (Brennan, 1996).

Most *Ribes* species produce perfect flowers, meaning that they have both male (anther) and female (stigma) organs. If these crops were entirely self-sterile, emasculation would not be required. However, at least partial self-fertility is the rule in black currants, and most red and white currants and gooseberries exhibit high degrees of self-fertility. To prevent pollen from a flower fertilizing that same flower (selfing), the male organs are removed.

Emasculation is performed during the bud stage after the flower clusters or inflorescences have elongated and the flower buds have swollen but have not yet opened. Keep (1975) reports that removing the ends of the flower clusters improves fruit set on the remaining flowers. Doing so also shortens the flowering time for the clusters, which normally bloom over a period of weeks. Some workers prefer to use a sharp scalpel or a hook sharpened on both sides to cut around the calyx tube (petals and sepals) just above the ovary, thereby removing all of the sepals, petals, and anthers at one time. Brennan (1996) advised that fruit set is usually best when no more than five to six flowers per inflorescence are emasculated. When the buds are tightly clustered on the inflorescence, such as is typical with red and white currants, a pair of fine-tipped forceps can be used to remove excess flowers and strip away the sepals, petals, and anthers on the flowers to be pollinated. In field pollinations or where insects are not completely screened out, prevent unwanted pollinations by enclosing the emasculated flowers in a waxed paper or fine mesh bag that is then tied tightly around the stem. Pollination bags are commercially available.

From one to three days after emasculation, apply pollen from the donor plant to the stigmas of the emasculated flowers. A fine-tipped paintbrush can be used to transfer the pollen (Keep, 1975). Alternatively, you can touch the stigma with an anther from a newly opened flower. After pollination, return the protective bags to field plants. Apply only one type of pollen or pollen mixture to the flowers on a single cluster. Carefully label each pollinated plant (or cluster, if multiple pollen donors are used for a single maternal plant) with the cultivar name, species, or breeder's designation for both the maternal parent and pollen donor. List the maternal parent first. Enter the pollinations into a logbook.

Although pollen can be transferred directly to the stigmas from newly opened flowers, it can also be collected into small glass or metal containers. Shaking newly opened flowers over an open container dislodges the pollen from the anthers. Intact anthers from open flowers can also be placed into the pollen container and agitated to release the pollen. Pollen stored at 68 to 75°F (20 to 24°C) in diffuse light loses viability within 14 days. Storing the pollen inside an enclosed container with a desiccant (calcium chloride) can extend its viability to 40 days (Raincikova, 1967). Brennan (1996) reported that the ability to extend pollen life is cultivar or genotype dependent, but that pollen viability in a range of genotypes declined to low levels after being stored dry, in the dark, at 39°F (4°C) for two to three weeks. Collecting pollen into containers facilitates making large numbers of crosses and also allows pollen from several donors to be mixed. In some cases of wide crosses, such as 'Crandall' *(R. aureum)* × gooseberries, pollen mixtures are more likely to result in fertilization and seed set than pollen from a single donor (Tolmacev, 1940). Pollen mixtures have also been popular with Russian breeders when crossing American and European gooseberries to produce mildew-resistant offspring (Keep, 1975). Whenever possible, use the freshest pollen available.

After pollination, allow the fruits to ripen. Keep fruits on field plants bagged until harvested. When harvesting, take care to keep fruits from different crosses separate and labeled. Harvest fruits when they are fully ripe but before they begin to shrivel. Extract the seeds by placing the berries into a blender partially filled with water. Place about twice as much water as fruit into the blender jar and run the blender at medium speed for 15 to 45 seconds. Remove gooseberry seeds from their skins by hand before placing the seeds and adhering pulp through the blending process. Masking the blender blades with rubber tubing reduces damage to the seeds. Transfer the slurry to a clear glass or beaker and fill the container with tap water. Viable seeds are heavy and will sink to the bottom. Hollow seeds, pulp, and skins can be floated off by repeatedly filling the container, allowing the heavy seeds to settle for a few seconds, and pouring off the dirty water.

To prevent damage from blender blades, Dr. Chad Finn of the U.S. Department of Agriculture-Agricultural Research Service Horticultural Crops Laboratory (personal communication) recommends extracting seeds without blending by using pectinase. Mash the fruit with the pectinase, add enough water to make a slurry, and allow the enzymes to work for 12 to 24 hours. The seeds settle to the bottom of the container and can be sieved or decanted out as described previously. Many brands of pectinase are available, and those marketed to juice manufacturing firms are relatively inexpensive.

Air dry the seeds at room temperature and under subdued light for about one week. Transfer the seeds to labeled paper, plastic, or aluminum foil envelopes and place the envelopes inside a desiccator at −4°F (−20°C). Seeds remain viable for several years in frozen storage (Brennan, 1996).

## Seed Germination and Sowing

As discussed in Chapter 3, *Ribes* seeds normally require stratification at temperatures near freezing before they will germinate. If they are sown soon after extraction, however, black currant seeds usually germinate within one to two weeks (Keep, 1975). When black currant seeds are sown without chilling, expose them to alternating daily temperatures of 75 to 86°F (24 to 30°C), followed by 8 to 12 hours at 39 to 59°F (4 to 15°C). Most breeders find it more convenient, however, to place the seeds inside plastic bags filled with moist sand or to sow the seeds in moist potting soil in seed flats placed inside a refrigerator or other cold location at about 37°F (3°C) for three to four months. Inspect the bags and seed flats regularly to ensure that the seeds do not become dry. Some *Ribes* species germinate at cold temperatures. If germination begins during stratification, move the seeds into a greenhouse or other growing location.

Seedlings are usually grown in a cold frame or greenhouse for about three months after developing two true leaves. After that, they are transplanted to field evaluation plots where they normally flower and bear fruit in one to two years' time (Keep, 1975). Brennan (1996) suggests first planting seedlings at one-foot (0.3 m) spacings in nursery rows. During the first two years, the plants are evaluated for their growth habits, disease resistance, and (for gooseberries) spininess. Those seedlings showing promise are then transplanted to three-foot (1 m) spacings for an additional four years or more to evaluate fruit and vegetative characteristics. Take great care to ensure all container- and field-grown plants are correctly labeled as to maternal parents and pollen donors. Observe seedlings regularly and maintain entries in a breeder's logbook.

Conventional breeding usually requires prolonged field trials to evaluate seedlings' fruit characteristics, resistance to diseases, cold hardiness, and other traits. Today, seedlings being screened for disease resistance are exposed to controlled inoculations of pathogens. This strategy allows for more rapid and accurate evaluations, and provides the opportunity to test resistance against multiple, known strains of a pathogen. Resistance to insect and mite pests is evaluated in much the same way.

Yields, plant habit, adaptability to temperature extremes, and similar characteristics require more lengthy field evaluations than do pest and disease resistance. Because plants respond differently under different growing conditions, evaluations of genetic and environmental (GXE) interactions require that potential cultivars be evaluated over a period of years at multiple, diverse sites.

Karnatz (1969) described a method of forcing black currant growth using elevated temperatures and long days to obtain fruits one year after sowing seed. In this study, black currant seeds were stratified at 3°C for eight weeks before being sown in pots in February. The pots were placed into a greenhouse set to 15°C day and 20°C night temperatures under 18-hour photoperiods, with supplemental lighting provided by lamps. In autumn (apparently mid-October), the growth area temperature was lowered to 5°C long enough for the plants to accumulate 50 hours of chilling. The temperatures were then increased to 25°C day and 10°C night with 18-hour photoperiods and supplemental lighting. New shoots were observed in about three weeks and flowers formed from early December through early January. The seedlings produced from 1 to 17 berries per plant in February, 12 months after sowing.

Accelerated growth may offer some advantages in screening seedlings for disease resistance or incorporation of a particular gene or set of genes, but caution is also required. The plants that formed in Karnatz's trials developed abnormally, with most new growth occurring at the branch ends and little shoot growth at the middle or bases of the canes. Berry production was too low to evaluate fruit quality, and evaluations of cane vigor, plant habit, yields, suitability for mechanical harvesting, and other important criteria were not possible. Long-term effects of early forcing on the plants were not determined, and Karnatz (1969) concluded that the results from the preliminary study did not clearly demonstrate that accelerated growth would be beneficial for practical breeding programs.

Decades often pass between the time a cross is made and the release of a cultivar. Nearly always, only a tiny percentage of seedlings produced in a cross prove worthy of naming and cultivation. Nor does a single cross always produce an acceptable cultivar. More often, cultivars are developed through crosses to incorporate a desirable gene, followed by repeated backcrosses to one of the parents to regain that original cultivar's characteristics. Throughout the evaluation process, it is important to compare the seedlings with standard cultivars whose characteristics are known and are growing at the same site and under the same conditions.

Breeding, even on a small scale, can quickly produce unmanageable numbers of seedlings. Be ruthless in rouging out undesirable plants as quickly as you can make a thorough evaluation. Unlike fine wines, disease-prone or otherwise undesirable plants do not improve with age and tie up land and a breeder's time. Keep only those plants that are improvements over their parents with regard to the gene or genes of interest, bearing in mind that multiple crosses and backcrossing may be necessary to produce commercially acceptable cultivars.

## Impacts of Self-Fertility and Self-Sterility on Breeding

Keep (1975) summarizes many evaluations of self-fertility and self-sterility in *Ribes*. Of the approximately 150 *Ribes* species worldwide, relatively few have been assessed for self-fertility. Based on results to date, however, self-sterility (self-incompatibility) predominates in *Ribes* species. In other words, pollinating flowers on an individual plant with pollen from that same plant typically results in little or no fruit set. Of the 21 species discussed by Keep (1975), *R. fontaneum* was self-fertile, *R. gracile* was partially self-fertile, and mixed results were reported for *R. dikuscha*. Unlike their wild cousins, currant and gooseberry cultivars seldom exhibit complete self-sterility, although cross-pollination often improves fruit set.

Self-fertility and self-sterility are important to breeders for several reasons. Backcrossing a genotype to itself or crossing to a closely related genotype can be difficult or impossible when strong self-sterility is present. Even should an $F_1$ offspring result from a cross, self-sterility remains a concern should further backcrosses be desirable. Furthermore, high degrees of self-fertility are desirable in cultivars. Partial or total self-sterility require that pollinizing cultivars be planted, complicating field layout, harvesting, and marketing. Even with pollinizing cultivars nearby, fruit set and yields can be reduced in strongly self-sterile cultivars.

### Inbreeding

Inbreeding depression is well known in many plant and animal species. *Ribes* are not exempt from inbreeding problems. Black currants, particularly western European cultivars, appear especially susceptible to inbreeding depression (Keep, 1975), which occurs when various combinations of deleterious and lethal genes are created during selfing and backcrosses. Kronenberg and Hofman (1965) found that selfed progenies of 'Baldwin', 'Daniels' September', 'Seabrook's Black', and 'Wellington XXX' exhibited decreased self-fertility and increased susceptibility to diseases, especially powdery mildew. Spinks (1947) noted reduced yields in selfed progenies of 'Baldwin', 'Boskoop', and 'French [Black]', concluding that selfing well-known cultivars was unlikely to produce outstanding plants.

The picture is not entirely clear, however, because offspring from selfed 'Consort' (*R. nigrum* × *R. ussuriense*) did not exhibit inbreeding depression (Kronenberg and Hofman, 1965). Likewise, Anderson and Fordyce (reported by Wood, 1960) found no effects of inbreeding on $S_1$ (first generation from a self-pollination) offspring of the Finnish cultivar 'Brodtorp'. Wilson (1970) conducted a detailed study on heterozygosity in black currants using 'Amos Black', 'Malvern Cross', 'Mendip Cross', and 'Wellington XXX'. These cultivars provided Wilson the opportunity to examine inbreeding depression on plants, depending on whether they arose from selfing, crosses between siblings, or crosses between half-siblings. In general, increasing heterozygosity resulted in greater bush vigor, longer internodes, longer strigs, and larger fruit. Selfing reduced cane vigor and regeneration (cane regrowth in the spring). Mildew resistance was not affected, although results suggested that susceptibility to leafspot increased with inbreeding.

Inbreeding problems develop most often while backcrossing. For example, an otherwise acceptable cultivar might be crossed with a genotype showing resistance to powdery mildew. Should some of the $F_1$ offspring inherit the mildew resistance, they are also likely to lose some of the desirable characteristics of the original cultivar. To restore lost characteristics, fruit breeders typically backcross $F_1$ plants and their offspring with original cultivars. In *Ribes* breeding, backcrossing can create problems, as discussed earlier. For example, Wilson (1970, p. 239) found with black currants that "[w]hile some of the consequences of inbreeding seemed conducive to high yields, others were apparently antagonistic, and it is suggested that the relationship between the parents should be taken more into account when planning a breeding program." Although less well documented, loss of vigor and other inbreeding problems occur in red currants, white currants, and gooseberries, much as they do in black currants (Keep, 1975). During backcrossing, breeders should use a wide range of cultivars as recurrent parents, rather than a single cultivar or a few closely related cultivars, to avoid inbreeding depression (Knight, 1983). Careful selection of parents based on their pedigrees, as well as their apparent characteristics, and detailed record keeping of crosses help reduce inbreeding difficulties.

*Crosses Between Species*

Many modern currant and gooseberry cultivars have been developed from more than one species. This trend for increased genetic diversity is likely to continue as traits for pest and disease resistance, long strigs, late flowering, and other desirable characteristics are identified in undomesticated *Ribes* species and incorporated into breeding programs. Not all crosses between species produce viable offspring, however. Keep (1962, 1975) discussed interspecific hybridization in detail. Generally speaking, cross compatibility within a single species is, with very few exceptions, the rule. Cross compatibility between morphologically (physically) similar species is also usual. Using taxonomic classifications by Berger (1924) and Rehder (1954), Keep (1962) created a classification that groups 53 *Ribes* species according to cross compatibility (Table 12.1). This classification is of great benefit to breeders, allowing them to focus on donors most likely to produce viable offspring.

Table 12.1 is divided into subgenera, sections, and species. Crosses made between species located within the same section generally produce interspecific hybrids that are at least partially fertile. According to Keep (1975, p. 214), however,

> [c]rosses between species of different sections or subgenera are often less successful, barriers to hybridization being evident at all stages, including initial failure to set fruit, failure of seed germination, death of the young seedlings (as in certain *R. nigrum* × *R. sanguineum* and *R. grossularia* × *R. sanguineum* combinations) and failure to flower. Finally, such hybrids as can be raised are usually sterile.

Keep did record 33 fertile intersectional hybrids. Keep (1975) observed that red currant × black currant and gooseberry × black currant hybrids are generally easy to produce and are often vigorous, but invariably sterile. Jostaberries were developed from sterile gooseberry × black currant hybrids that were induced to form fertile allotetraploids by the application of colchicine. Although more complicated than breeding within species and with closely related species, intra- and intersectional crosses provide opportunities to introduce valuable pest- and disease-resistance genes into domestic crops (Bauer, 1955).

TABLE 12.1. Taxonomy of *Ribes* species based on cross compatibility.

| Subgenus | Section | Species | Distribution |
|---|---|---|---|
| *Bersia* Spach | *Euberisia* Jancz. | *R. acuminatum* Wall. | Asia |
| | | *R. alpinum* L. | Europe |
| | | *R. glaciale* Wall. | Asia |
| | | *R. luridum* Hook. | Asia |
| | | *R. maximowiczii* Batal. | Asia |
| | | *R. orientale* Desf. | East Europe, North Asia |
| | *Diacantha* Jancz. | *R. dicanthum* Pall. | Asia |
| | *Davidia* Jancz. | *R. laurifolium* Jancz. | Asia |
| | *Hemibotrya* Jancz. | *R. fasciculatum* Sieb. & Zucc. | East Asia |
| | *Parilla* Jancz. | *R. gayanum* (Spach) Steud. | South America |
| | | *R. integrifolium* Phil. | South America |
| | | *R. polyanthes* Phil. | South America |
| | | *R. punctatum* Ru. & Pav. | South America |
| | | *R. valdivianum* Phil. | South America |
| *Microsperma* Jancz. | | *R. ambiguum* Maxim. | Asia |
| *Symphocalyx* Berl. | | *R. aureum* Pursh. (*R. odoratum* Wendl.) | North America |
| *Calobotrya* Spach | *Calobotrya* Spach | *R. ciliatum* H. & B. | North America |
| | | *R. glutinosum* Benth. | North America |
| | | *R. malvaceum* Sm. | North America |
| | | *R. nevadense* Kellogg | North America |
| | | *R. sanguineum* Pursh. | North America |

TABLE 12.1 *(continued)*

| Subgenus | Section | Species | Distribution |
|---|---|---|---|
| | *Eucoreosma* Jancz. | *R. americanum* Mill. | North America |
| | | *R. bracteosum* Dougl. | North America |
| | | *R. dikuscha* Fisch. | North Asia |
| | | *R. hudsonianum* Rich. | North America |
| | | *R. nigrum* L. | Europe, Central Asia |
| | | *R. ussuriense* Jancz. | East Asia |
| *Cerophyllum* Spach | | *R. cereum* Dougl. | North America |
| *Heritiera* Jancz. | | *R. laxiflorum* Pursh. | North America |
| *Ribesia* Berl. | | *R. longeracemosum* Franch. | Asia |
| | | *R. multiflorum* Kit. | East Europe |
| | | *R. petraeum* Wulf. | Europe |
| | | *R. rubrum* L. | Europe, North Asia |
| | | *R. sativum* Syme | West Europe |
| | | *R. warscewiczii* Jancz. | North Asia |
| *Grossularioides* Jancz. | *Grossularioides* | *R. lacustre* (Pers.) Poir. | North America |
| | | *R. montigenum* McClatchie | North America |
| *Grossularia* Rich. | *Eugrossularia* Engl. | *R. aciculare* Sm. | North Asia |
| | | *R. cynobasti* L. | North America |
| | | *R. divaricatum* Dougl. | North America |
| | | *R. grossularia* L. | Europe, North Africa |
| | | *R. grossularioides* Maxim. | East Asia |
| | | *R. hirtellum* Michx. | North America |

| Subgenus | Section | Species | Distribution |
|----------|---------|---------|--------------|
| | | *R. inerme* Rydb. | North America |
| | | *R. irriguum* Dougl. | North America |
| | | *R. missouriense* Nutt. | North America |
| | | *R. niveum* Lindl. | North America |
| | | *R. oxyacanthoides* L. | North America |
| | | *R. pinetorum* Greene | North America |
| | | *R. rotundifolium* Michx. | North America |
| | | *R. stenocarpum* Maxim. | Asia |
| | *Robsonia* Bert. | *R. lobbii* Gray | North America |
| | | *R. menziesii* Pursh. | North America |

*Source:* Adapted from Rehder (1954), Keep (1962), and Brennan (personal communication, 2002).
*Note:* Intrasectional hybrids should normally be fertile and intrasubgeneric hybrids at least partially fertile. Those intersubgeneric hybrids that can be raised to maturity will normally be sterile. Within subgenus *Calobotrya*, production of fertile hybrids through intersectional crosses between sections *Calobotrya* and *Eucoreosma* is possible, although not necessarily easy.

Readers interested in more detailed discussions of *Ribes* genetics, breeding, and cytology are referred to Keep (1962, 1975) and Brennan (1990, 1996).

## Pollen and Plant Treatment Used by Breeders

Keep (1975) described numerous attempts by researchers to manipulate fertilization and fruit set by treating pollen with ultraviolet light, gamma radiation, X-rays, and giberellic acid (GA). These treatments have been used in attempts to overcome self-incompatibility, improve fertilization and fruit set in wide crosses, and increase seed set in inbred lines. Ultraviolet radiation allowed for selfed seed in the normally self-incompatible *R. aureum* and *R. bracteosum,* but failed to do so with *R. sanguineum, R. multiflorum,* and *R. longeracemosum* (Arasu, 1968). Gamma irradiation of pollen improved seed set in a black currant × *R. americanum* cross and enhanced the germination of seeds from black currant × red currant crosses (Melekhina, 1968). Applications of 0.02 to 0.03 percent GA to the stigmas improved seed set in black currant × red currant × *R. aureum* crosses and black currant × white currant crosses (Cuvasina, 1961, 1962).

The chemical colchicine has been used by *Ribes* breeders, particularly when trying to produce fertile allotetraploids (interspecific hybrids having four sets of chromosomes). This chemical works by disrupting the cell division process to produce plants (sometimes called colchiploids) that have more sets of chromosomes than normal. Jostaberries (tetraploids) were developed using colchicine. The two extra sets of chromosomes allow these normally sterile hybrids to produce fruit. Mature *Ribes* plants are quite resistant to colchicine, requiring long exposures and repeated applications for the chemical to take effect. Young seedlings and germinating seeds are more responsive than older plants to colchicine treatments (Keep, 1975). Colchicine is not without its drawbacks, however, and can cause mutations in plants and animals alike. It is classified as a poison and must be handled with great care.

## Modern Ribes Breeding

Technological advances in plant breeding have revolutionized *Ribes* cultivar development. Work at the Scottish Crop Research Institute and other advanced breeding facilities now involves identification of specific genes or clusters of genes that control desirable traits. This ability creates the potential to incorporate new characteristics into otherwise acceptable cultivars without the need for extensive crosses and backcrosses. Similar advances allow researchers to develop markers for valuable sections of DNA that control or correspond to desirable traits. When such markers have been identified, breeders, using a DNA profiling technique called marker-assisted selection (MAS), can screen large numbers of very young seedlings or genotypes to identify those that possess a desired trait without the need for lengthy field trials.

As discussed earlier, careful selection of parents is extremely important in a successful breeding program. Based on germplasm evaluations and breeding efforts in Europe and North America, recommendations have been developed regarding donors for specific traits.

Suggested donors are listed in Table 12.2. Bear in mind that these recommendations are not exhaustive and other donors are certainly possible. The tables should, however, serve as a starting point for *Ribes* breeding efforts in North America.

Given the large numbers of species and cultivars available throughout the world, relatively open borders for germplasm exchanges, a detailed knowledge of *Ribes* genetics, and widely available technical information, prospects for developing high-quality currant, gooseberry, and jostaberry cultivars adapted to North America are excellent.

TABLE 12.2. Suggested parents to obtain desired qualities in *Ribes*.

| Quality | Black currant | Red and white currant | Gooseberry | Species |
|---|---|---|---|---|
| High yields | Ben Alder<br>Ben Lomond<br>Ben Nevis<br>Ben Sarek[a]<br>Brodtorp<br>Coronet | Earliest of Fourlands<br>Heinemanns Rote Spatl.<br>Jonkheer van Tets<br>Red Lake<br>Rondom | Careless<br>Downing<br>Green Gem<br>Houghton<br>Izumrud<br>Keepsake<br>Lancashire Lad<br>Malakhit<br>Pioner<br>Poorman<br>Russkij<br>Smena<br>Whinham's Industry<br>Whitesmith | *R. dikuscha*<br>*R. divaricatum*<br>*R. longeracemosum*<br>*R. missouriense*<br>*R. multiflorum*<br>*R. nigrum sibiricum*<br>*R. oxyacanthoides*<br>*R. succirubrum* |
| Large fruit size | Ben Lomond<br>Ben Sarek<br>Black September<br>Boskoop Giant<br>Victoria<br>Goliath<br>Strata | Cherry<br>Fay's Prolific<br>Heros<br>Jonkheer van Tets<br>Laxton's Perfection<br>Rosa Sudmark<br>Rote Kirsch<br>Versailles<br>Weisse aus Juterbog | Antagonist<br>Blood Hound<br>Broom Girl<br>Careless<br>Dan's Mistake<br>Green Ocean<br>Grune Flaschenbeere<br>Grune Edelstein<br>Gunner<br>Invicta<br>Jumbo<br>Keepsake<br>Lord Derby<br>Maurer's Seedling<br>Telegraph<br>Thumper | *R. aureum*<br>*R. nigrum*<br>*R. nigrum sibiricum*<br>*R. macrocarpon*<br>*R. petraeum atropurpur* |

| Objective | Cultivars | | | Ribes species |
| --- | --- | --- | --- | --- |
| Tough fruit skin to facilitate mechanical harvesting | Baldwin<br>Ben Lomond<br>Black Reward<br>Invigo | Gondouin<br>Heinemanns Rote Spat.<br>Rode Komeet<br>Rondom | All reasonably firm | Unknown |
| Easy strig removal to facilitate mechanical harvesting | Bang Up<br>Black Reward<br>Goliath<br>Invigo<br>Roodknop<br>Seabrook's Black<br>Topsy | Rondom | Captivator<br>Fredonia<br>Glenashton<br>Silvia | *R. bracteosum*<br>*R. glutinosum*<br>*R. longeracemosum*<br>*R. sanguineum* |
| Long, easy-to-remove strigs to facilitate hand harvesting | Boskoop Giant<br>Jet<br>Invincible Giant (Prolific)<br>Westwick Choice | Heinemanns Rote Spatl.<br>Red Lake<br>Rondom | Not applicable | *R. bracteosum*<br>*R. longeracemosum*<br>*R. multiflorum* |
| High ascorbic acid (vitamin C) content | Altajskaja Deserthaja<br>Baldwin<br>Ben Connan<br>Ben Lomond<br>Ben Tirran<br>Golubka<br>Koksa<br>Moskovskaya<br>Pobyeda<br>Stakhanovka Altaya<br>Tisel | Not as high as black currant | Roman | *R. aureum*<br>*R. dikuscha*<br>*R. divaricatum*<br>*R. nigrum sibiricum*<br>*R. pauciflorum*<br>*R. petraeum atropurpureum*<br>*R. succirubrum* |

TABLE 12.2 (continued)

| Quality | Black currant | Red and white currant | Gooseberry | Species |
|---|---|---|---|---|
| Improved flavor | Baldwin<br>Brodtorp<br>Crandall<br>Goliath<br>Noir de Bourgogne<br>Royal de Naples<br>Silvergieters Zwarte<br>Swedish Black | Hoornse Rode<br>Laxton's No. 1<br>Laxton's Perfection<br>Fays Prolific | Glenton Green<br>Golden Drop<br>Green Walnut<br>Hoennings Fruheste<br>Howard's Lancer<br>Ironmonger<br>Langley Gage<br>Leveller<br>Pitmaston Greengage<br>Red Champagne<br>Roseberry<br>Scotch Red Rough<br>Warrington<br>Whinham's Industry<br>Whitesmith<br>Yellow Champagne | *R. grossularia*<br>*R. nigrum* |
| High anthocyanin and phenolic concentrations | Belorusskaja Slodkaja<br>Ben Alder<br>Ben Lomond<br>Consort<br>Crusader<br>Janslunda<br>Lunnaja<br>Silvergieters Zwarte<br>Willoughby | Low compared with black currants | Low compared with black currants; includes dark-pigmented gooseberry cultivars | *R. cynobasti*<br>*R. nigrum*<br>*R. niveum*<br>*R. pauciflorum*<br>*R. robustum*<br>*R. valdivianum* |

| Erect growth habit | Consort<br>Goliath<br>Moskovskaya<br>Narjadnaja<br>Pamjat Micurina<br>Pobyeda<br>Westra | Most cultivars are erect | Careless<br>Greenfinch<br>Izumrud<br>Reverta<br>Ristula<br>Risulfa<br>Rokula<br>Rubin<br>Smena<br>Whinham's Industry<br>Whitesmith | *R. alpestre*<br>*R. leptanthum*<br>*R. niveum*<br>*R. sanguineum*<br>*R. watsonianum* |
|---|---|---|---|---|
| Improved ease of propagation | Generally not a problem | Generally not a problem | Houghton<br>Jubilenjyi<br>Russkij<br>Smena | *R. divaricatum*<br>*R. hirtellum*<br>*R. oxyacanthoides*<br>*R. succirubrum* |
| Late flowering or increased spring frost resistance | Ben Alder<br>Ben Lomond<br>Ben Nevis<br>Ben Tirran<br>Brodtorp<br>French Black<br>Goliath<br>Jet<br>Ojebyn<br>Pilot A. Mamkin | Heinemanns Rote Spat.<br>Kernlose<br>Moore's Ruby<br>Prince Albert<br>Raby Castle<br>Stanza<br>Victoria | Broom Girl<br>Captivator<br>Houghton<br>Krasnyi Krupny<br>Lancer<br>Lord Derby<br>Resistenta<br>Robustenta<br>Trumpeter | *R. bracteosum*<br>*R. dikuscha*<br>*R. divaricatum*<br>*R. mandshuricum*<br>*R. multiflorum*<br>*R. nigrum*<br>*R. nigrum sibiricum*<br>*R. petraeum* |

TABLE 12.2 (continued)

| Quality | Black currant | Red and white currant | Gooseberry | Species |
|---|---|---|---|---|
| Increased winterhardiness | Brodtorp<br>Golubka<br>Imandra<br>Lepaan Musta<br>Ojebyn<br>Omsk<br>Pechora<br>Primorskij Cempion<br>Sunderbyn II | Houghton Castle<br>Kandalaksa<br>Komovaja Markina<br>Raby Castle<br>Red Dutch<br>Red Lake<br>Varzuga<br>Victoria<br>White Grape | Downing<br>Finland 1<br>Hankkijas Delikatess<br>Hinnonmakis Gula<br>Houghton<br>Malakhit<br>Pioner<br>Poorman<br>Russkij<br>Scania<br>Silvia | R. aciculare<br>R. burejense<br>R. cynobasti<br>R. dukuscha<br>R. divaricatum<br>R. gracile<br>R. hirtellum<br>R. nigrum sibiricum<br>R. pauciflorum<br>R. petraeum<br>R. procumbens<br>R. rotundifolium<br>R. rubrum<br>R. stenocarpum<br>R. triste |
| Reduced spines in gooseberries | Not applicable | Not applicable | Captivator<br>Houghton<br>Poorman<br>Spinefree<br>Welcome | R. cynobasti<br>R. hirtellum<br>R. inerme<br>R. missouriense<br>R. robustum |
| Low chilling requirement | Altajskaja Desertnaja<br>Baldwin<br>Blackdown<br>Cotswold Cross<br>Koksa<br>Zoja<br>Plancher<br>Dordrechter | Red Versailles 87<br>Rondom<br>Wilder | Unknown | Unknown |

| | | | |
|---|---|---|---|
| Drought tolerance | Desertnaja Altaya<br>Pamjat Micurina<br>Saunders | Unknown | Artemovskij<br>Bakhmut<br>Donetskij<br>Krupnoplodnyi<br>Donetskij Pervenets<br>Yubilenyi | R. aureum |
| Resistance to wind damage | Not a serious problem | Ayrshire Queen<br>Earliest of Fourlands<br>Jonkheer van Tets<br>New Red Dutch<br>Red Lake<br>Prince Albert<br>Victoria | Not a serious problem | Unknown |
| Improved tolerance to sulfur fungicides | Unknown | Unknown | May Duke<br>Whinham's Industry | Unknown |
| Resistance to American powdery mildew | Ben Alder<br>Ben Sarek<br>Ben Tirran<br>Brodtorp<br>Crandall Micurina<br>Karelskaya<br>Lepaan Musta<br>Matkakowski<br>Ojebyn<br>Primorskij Cempion<br>Stor Klas<br>Sunderbyn II<br>Titon<br>Troll | Fay's Prolific<br>Houghton Castle<br>Jonkheer van Tets<br>Minnesota 71<br>Red Lake<br>Redstart<br>Rondom<br>Versailles<br>Wentworth Leviathan<br>White Transparent | Captivator<br>Dr. Tornmark<br>Hankkijan Herkku<br>Hankkijas Delikatess<br>Hinnonmaen Keltainen<br>Hinnonmakis Gula<br>Houghton<br>Izumrud<br>Malakhit<br>Pellervo<br>Plodorodnyj<br>Rekord<br>Remarka<br>Resistenta<br>Reverta<br>Ristula<br>Risulfa<br>Robusta<br>Rokula<br>Russkij | R. alpinum<br>R. americanum<br>R. aureum<br>R. cereum<br>R. cynobasti<br>R. dikuscha<br>R. divaricatum<br>R. glutinosum<br>R. hirtellum<br>R. hudsonianum<br>R. Irriguum<br>R. janczewskii<br>R. leptanthum<br>R. longeracemosum<br>R. multiflorum<br>R. nigrum sibiricum<br>R. niveum<br>R. oxyacanthoides<br>R. pauciflorum<br>R. petiolare |

TABLE 12.2 *(continued)*

| Quality | Black currant | Red and white currant | Gooseberry | Species |
|---|---|---|---|---|
| Resistance to American powdery mildew *(continued)* | | | Sabine | *R. petraeum* |
| | | | Scania | *R. sanguineum* |
| | | | Sebastian | *R. warscewiczii* |
| | | | Shefford | *R. watsonianum* |
| | | | Smena | |
| | | | Stanbridge | |
| | | | Sutton | |
| | | | Rixanta | |
| | | | Reflamba | |
| | | | Rolanda | |
| Resistance to white pine blister rust | Bzura | Birnformige Weisse | Careless[b] | *R. aciculare* |
| | Consort[b] | Earliest of Fourlands | Clark[b] | *R. alpinum* |
| | Coronet[b] | Eyath Nova | Columbus[b] | *R. americanum* |
| | Crandall[b] | Franco-German | Crown Bob[b] | *R. cereum* |
| | Crusader[b] | Gloire de Sablons[b] | D. Young[b] | *R. cynobasti* |
| | Doez Sibirjoczk[b] | Gogingers Birnform | Downing[b] | *R. dicanthum* |
| | Dunajec | Heros | Early Sulfur[b] | *R. dikuscha* |
| | Loda | Holland Redpath | Glenton Green[b] | *R. glaciale* |
| | Lowes Auslese[b] | Houghton Castle | Golda[b] | *R. glutinosum* |
| | Lumnaja[b] | London Market[b] | Greenfinch[b] | *R. hallii* |
| | Ner | New York 72[b] | Hinnonmaen Keltainen[b] | *R. innominatum* |
| | Odra | Red Dutch | Hoennings Frueheste[b] | *R. leptanthum* |
| | Polar[b] | Rivers | Howard's Lancer[b] | *R. lurudum* |
| | Primorskij Cempion | Rondom[b] | Industry[b] | *R. nigrum sibiricum* |
| | Rain-in-the-face[b] | Rote Kernlose | Invicta | *R. orientale* |
| | Sligo[b] | Sabine[b] | Josselyn[b] | *R. pauciflorum* |
| | Titania[b] | Simcoe King | Jumbo[b] | *R. petraeum* |
| | Warta | Victoria | Poillaji dindis[b] | *R. pinetorum* |
| | Wista | Viking[b] | Robustenta[b] | *R. procumbens* |
| | Willoughby[b] | Weisse Burgdorfer | Sabine | *R. rubrum* |
| | | | Schultz[b] | *R. ussuriense* |
| | | | Told Gyostes[b] | *R. viburnifolium* |
| | | | Whitesmith[b] | |
| | | | White Lion[b] | |
| | | | Weisse Votragende[b] | |

| | | | | |
|---|---|---|---|---|
| Resistance to Anthracnose leafspot | Barhatnaja<br>Belorusskaja Pozdnjaja<br>Boskoop Giant<br>Brodtorp<br>Gerby<br>Goliath<br>Izbrannaja<br>Junnat<br>Kaskad<br>Laxton's Raven<br>Losickaja<br>Mendip Cross<br>Metepa<br>Minskaja<br>Nahodka<br>Neosypajuscajasja<br>Ojebyn<br>Podmoskovnaja<br>Seabrook's Black<br>Sopiernik<br>Uspekh<br>Victoria<br>Westwick Triumph | Earliest of Fourlands<br>Heinemann's Rote Spat.<br>Prince Albert<br>Red Dutch<br>Rondom<br>Viking | Belorusskij<br>Captivator<br>Como<br>Izjumny<br>Jarovoj<br>Krasavec Losicy<br>Malakhit<br>Ogonek<br>Pioner<br>Pixwell<br>Plodorodnyj<br>Remarka<br>Resistenta<br>Rideau<br>Robustenta<br>Russkij<br>Scedryj<br>Transparent | R. americanum<br>R. aureum<br>R. burejense<br>R. ciliatum<br>R. cynobasti<br>R. dukuscha<br>R. divaricatum<br>R. glutinosum<br>R. gracile<br>R. irriguum<br>R. moupinense<br>R. multiflorum<br>R. nigrum sibiricum<br>R. niveum<br>R. non-scriptum<br>R. oxyacanthoides<br>R. pauciflorum<br>R. petraeum<br>R. pubescens<br>R. rotundifolium<br>R. rubrum<br>R. sanguineum<br>R. warscewiczii<br>R. watsonianum |
| Resistance to Septoria leafspot | Barhatnaja<br>Barnaulka<br>Belorusskaja Pozdnjaja<br>Bija<br>Golubka<br>Irtysh<br>Kantata<br>Karelskaya<br>Lana<br>Nina<br>Pamjat Micurina<br>Primorskij Cempion<br>Stakhanovka Altaya | Red Cross<br>Red Dutch | Malakhit<br>Pioner<br>Plodorodnyj<br>Resistenta<br>Robustenta<br>Russkij | R. anericanum<br>R. dikuscha<br>R. divaricatum<br>R. nigrum sibiricum<br>R. pauciflorum |

225

TABLE 12.2 *(continued)*

| Quality | Black currant | Red and white currant | Gooseberry | Species |
|---|---|---|---|---|
| Resistance to Botryosphaeria cane blight | Crandall | Fox New<br>Knights Sun Red<br>La Constante<br>Littlecroft<br>Minnesota 77<br>Perfection<br>Prince Albert<br>Red Dutch<br>Stephens No. 9<br>White Dutch | Unknown | Unknown |
| Resistance to Botrytis dieback | Bija<br>Ben Alder<br>Ben Tirran | Fay's Prolific<br>Hoornse Rode<br>Maarse's Prominent<br>Palandts Seedling<br>Stern des Nordens<br>Victoria | Carrie<br>Cernyi Negus<br>Houghton<br>Micurinskij<br>Mysovskij 17<br>Mysovskij 37<br>Pjatiletka | *R. americanum* |
| Resistance to black currant reversion virus[c] | Ben Gairn<br>Blestjascaja<br>Foxendown<br>Golubka<br>Novost<br>Pamjat Micurina<br>Stakhanovka Altaja<br>Uspekh | Unknown | Not applicable | *R. aureum*<br>*R. cereum*<br>*R. dikuscha*<br>*R. longeracemosum*<br>*R. nigrum sibiricum*<br>*R. niveum* |
| Resistance to vein banding virus | Most cultivars tolerant | None immune<br>Rondom (tolerant) | Mountain Seedling<br>Oakmere | *R. divaricatum*<br>*R. sanguineum* |

| | | | |
|---|---|---|---|
| Resistance to leaf-curling midge (*Dasyneura tetensi* Rubs.)[d] | Dikovinka<br>Golubka Seedling<br>Kangosfors<br>Koksa<br>Minaj Smirev<br>Partizanka<br>Primorskij Cempion<br>Sunderbyn II<br>Zoja | Generally not susceptible | Generally not susceptible | *R. alpinum*<br>*R. americanum*<br>*R. aureum*<br>*R. cereum*<br>*R. ciliatum*<br>*R. dikuscha*<br>*R. divaricatum*<br>*R. glutinosum*<br>*R. grossularia*<br>*R. longeracemosum*<br>*R. orientale*<br>*R. sanguineum* |
| Resistance to black currant gall mite[e] | Ben Hope<br>Foxendown<br>Narjadnaja<br>Neosypajuscajasja<br>Primorskij Cempion<br>Seabrook's Black<br>Stakhanovka Altaya | Unknown | Gooseberries are resistant to the mite | *R. cereum*<br>*R. glutinosum*<br>*R. grossularia*<br>*R. janczewskii*<br>*R. nigrum sibiricum*<br>*R. pauciflorum*<br>*R. ussuriense* |

[a]Highest yielding of the 'Ben' series.
[b]Highly resistant or immune after artificial inoculation in Corvallis, Oregon.
[c]Black currant reversion virus is not presently known in North America but exists in Europe, New Zealand, and Australia.
[d]The pest *Dasyneura tetensi* Rubs., which creates problems in black currant nurseries, is not known in North America but is present in Europe.
[e]Black currant gall mite, which carries black currant reversion disease, is not presently known in North America but is established in Europe, New Zealand, and Australia.

# Appendix A

# Site Selection Checklist

The following checklist is provided for evaluating potential farming operations and sites. The checklist should be used only after carefully reading and considering the preceding text.

## NONSITE CONSIDERATIONS

I. The product
    A. What do you want to grow?
    B. Do you know the climatic and cultural needs of currants, gooseberries, and jostaberries?
    C. Do you plan to sell your fruit fresh or for processing?
    D. If your market is processing, will you
        1. Ship your berries to a processor? If so, what processor?
        2. Create your own value-added product?

II. Available resources
    A. Personal skills
        1. Do you have necessary production and marketing skills?
        2. Will you need to hire a manager?
        3. Is a qualified manager available?
    B. Financial resources
        1. Land
            a. Do you now own or rent a suitable farm site?
            b. How much does it cost you to own or rent the site?
            c. How much will it cost to purchase or rent a site?
            d. Will site improvements be needed?
            e. How much will site improvements cost?
        2. Facilities
            a. What facilities will your berry farm require?
            b. Can you construct some of your own facilities?
            c. What will the required facilities cost?

    3. Capital
       a. Do you have the necessary capital?
       b. Can you obtain sufficient capital?
       c. How much will obtaining capital cost you?
       d. What financial rate of return do you expect from your enterprise?

III. Marketing the product
    A. Where is your market and how will you get your product there?
       1. Roadside stand
         a. Is there sufficient traffic of people who will buy your fruit and other products?
         b. Is your stand easily visible?
         c. Are there easy access and parking?
         d. Are the surroundings conducive to marketing?
         e. Can you place advertising signs along the road?
       2. Local direct sales
         a. Is a large enough population nearby?
         b. Will local businesses purchase your products?
       3. Export sales
         a. Is an export market open to you?
         b. Can you support an export business?
         c. What is your access to truck and air transport?
         d. How much will it cost you to export your fruit or value-added products?

IV. Labor
    A. Labor needs, resources, and costs
       1. How much labor do you need and at what times of the year?
       2. Will workers come from local or migrant sources?
       3. Is a sufficient labor pool available to meet your needs?
       4. What will the cost of labor be?
    B. Meeting legal requirements
       1. Can you meet all legal requirements for labor?
       2. Are sanitary facilities available?
       3. Is shelter available?
       4. Is adequate and legal housing available?
       5. Can you meet labor transportation needs?

V. Support services
    A. Can you form a cooperative with other farmers?
    B. Are equipment and supply facilities nearby?
    C. Are processors available nearby?
    D. Is a commercial kitchen available to you?
    E. Are production and marketing technical support available?

## *SITE CONSIDERATIONS*

I. Climate
   A. Is the growing season at your site at least 120 to 150 days long?
   B. What minimum winter temperatures can you expect?
   C. What are the spring and fall frost dates?
   D. How much precipitation do you get, and when during the year do you get it?
   E. Do summer daytime temperatures remain above 85°F (29°C) for more than a few days at a time?

II. Soil
   A. What is the drainage and water-holding capacity?
   B. What is the soil pH and calcium content?
   C. What is the organic matter content?
   D. How deep is your topsoil, and what lies below it?
   E. What is the texture of your topsoil?

III. Water
   A. How much and what quality of water will you need?
   B. What are your water rights?
   C. Available sources of water
      1. Well—How deep? Capacity?
      2. Spring—Seasonal or year-round? Capacity?
      3. Stream or lake—Water rights? Seasonal?
      4. Farm pond—Capacity? Seasonal?
      5. How much will accessing water cost?
      6. Type of irrigation system desired
         a. Flood or furrow
         b. Overhead sprinkler
         c. Trickle or drip

IV. Topography
   A. What is the air drainage?
   B. What is the light exposure?
   C. Does the site have a north-facing slope?
   D. Will site temperatures be moderated by large bodies of water?
   E. Are air pollution sources nearby?
   F. What are nearby farming operations like?

V. Winds
   A. What are the direction, frequency, duration, and velocity of prevailing winds?

VI. *Ribes* pests and diseases endemic to the area
   A. What pest and disease problems should you anticipate?

VII. History of the site
    A. What crops or livestock were previously produced on the site?
    B. What was the previous pesticide use on the site?
    C. If you are interested in farming organically, will you be able to meet certification standards?
    D. Are toxic chemicals or buried storage tanks a problem?
VIII. Access and utilities
    A. Access
        1. Roads
            a. When will you need to use farm and access roads?
            b. What types of vehicles will be using the roads?
            c. Are access roads suitable for transporting your fruit?
            d. Will you have access whenever you need it?
            e. Who provides and pays for road maintenance?
            f. Will your operation adversely impact the roads?
            g. Are there other residences or businesses along your roads?
            h. Can you safely and economically transport your product to market?
        2. Is suitable parking available?
    B. Utilities
        1. Will electricity, natural gas, and telephones be needed?
        2. Are these utilities available at the site?
        3. How much will installation and utility use cost?
IX. Zoning, taxes, and ordinances
    A. Zoning
        1. Will county and city ordinances and zoning regulations allow you to operate your selected business?
        2. Does the government regulate your site?
            a. Highly erodible land?
            b. Designated wetlands?
    B. Taxes
        1. What taxes will be assessed at your location?
        2. Are tax incentives available?
    C. Annexation
        1. Is your site likely to be annexed?
    D. Liens and restrictive covenants
        1. Are there any liens or restrictive covenants on the property?

# Appendix B

# Enterprise Budget

The following tables show typical costs of producing currants and gooseberries for sale in the fresh market. Because jostaberries have not yet been produced on a commercial scale, no accurate production-cost estimates are available. Prospective jostaberry producers may find the gooseberry budget useful in evaluating their enterprises.

An established farmer who diversifies into raising currants, gooseberries, or jostaberries has great advantages over someone starting from scratch. Land and equipment are generally already available, and the farmer is experienced in commercial production. Established crops can provide a positive cash flow until the new crops come into production and are being marketed.

The figures here are based on typical cultural practices, although individual operations differ depending on the region, management style, and horticultural practices. The figures, given in U.S. dollars, are conservative in terms of maximizing costs, minimizing returns, estimating the time required for the bushes to reach harvest maturity, and potential yields. The assumptions that this budget is based on are as follows:

1. The *Ribes* enterprise covers two acres.
2. The land is owned by the farmer and has a value of $3,000 per acre. Annual property taxes are $30 per acre. Prior to growing currants and gooseberries, the land was used for dryland grazing.
3. For currants, year one is an establishment year. During years two and three, the bushes are maintained but are not harvested. Partial crops are harvested in years four and five, and full production is attained in year six.
4. For gooseberries, year one is an establishment year. During years two and three, the bushes are maintained but are not harvested. Partial crops are harvested in years four through six. Full production is attained in year seven.
5. Hired hourly labor is valued at $9.50 per hour, including workers' compensation, unemployment insurance, and other labor overhead expenses. The owner's labor is figured at $12.50 per hour.

6. The grower is starting from scratch and must purchase all equipment and structures. Equipment and structures are purchased new.
7. Establishment costs are amortized or spread out over the 20-year expected life of the planting. Interest is figured at 12 percent.
8. At full production, projected yields are listed here. Actual yields will depend on plant spacing, varieties, weather, and management practices.

| | |
|---|---|
| black currants | 5,500 lb/acre |
| red currants | 10,000 lb/acre |
| white currants | 12,000 lb/acre |
| gooseberries | 6,000 lb/acre |

## Establishment Costs

The cost of establishing the currant or gooseberry stand must be recovered over the stand's useful life. This process, called amortization, involves carrying forward, with interest, the total establishment costs for years zero, one, and two. The total establishment costs are amortized over an expected life of 20 years. The average interest is assumed to be 12 percent.

## Budgets

The two categories of costs listed in the budgets are fixed and variable costs. Variable costs are those over which you have direct control. They can be increased or decreased at your discretion, or avoided if you choose not to produce. Variable costs increase as the level of production increases. Examples of variable costs are planting stock, fertilizer, pesticides, fuel, repairs, hired labor, and interest on operating capital. Fixed costs are those that remain unchanged, no matter how much fruit you produce, or whether you produce any at all. They are associated with owning equipment, buildings, and other fixed inputs, including depreciation, taxes, insurance, and interest.

Fixed and variable costs can be either cash or noncash costs. Cash costs are out-of-pocket expenses. They can be variable, such as fuel, or fixed such as property taxes. Cash costs must be paid outright. Noncash costs do not involve an immediate cash payment. For example, when you provide your own labor, cash is not exchanged and your labor is considered a noncash cost. Unfortunately, too many beginning farmers forget to pay themselves and end up working for nothing. If you choose to hire labor for the same job, then the labor expenses are cash costs. Accounting for noncash costs is particularly important in analyzing the actual cost of an enterprise. For this reason, both cash and noncash costs are treated as expenses in the following budgets.

Long-term, intermediate, and short-term capital are used in these tables to finance establishment costs, equipment, permanent structures, irrigation, and operating inputs. Interest on operating capital is assumed to be 12 percent and is treated as a cash expense. Interest on investment is figured at 12 percent and is treated as a noncash expense. What this means is that if you had chosen to invest your money in something else, stocks for example, you would have received a projected return of 12 percent. Overhead accounts for 5 percent of each year's variable costs and includes expenses such as insurance, office supplies, and telephone bills.

## Using This Budget

This budget provides an estimate of the time and cost involved in starting a currant or gooseberry farm from scratch. Use it as a starting point for planning your own operation. For example, you may already own a tractor or other equipment, thereby reducing your machinery and establishment costs. Used equipment costs less initially than new equipment, but repair costs are usually higher. If you plan to have all of the fruit harvested by customers in a U-pick setting, you will save on harvest labor and refrigeration costs, but you must also account for increased advertising and reduced yields created by unharvested fruit. Mechanical harvesters reduce picking labor but increase equipment costs and may limit your market to processing fruit only. Using preemergence herbicides reduces hand-weeding costs, but will prevent selling the fruit at a potential premium on the organic market.

The far right columns of Tables B.8 through B.11 are blank, allowing you to customize the tables for your planned operation. In calculating your own estimates, be conservative. Avoid the temptation to underestimate costs and overestimate returns. When calculating labor costs, remember to pay yourself. The time you spend growing currants and gooseberries could be invested in other income-producing activities.

TABLE B.1. Yields and costs for red currant production.

| Year | Stage of production | Yield (lb/acre) | Cost/acre ($) | Cost/pound ($) |
|------|---------------------|-----------------|---------------|----------------|
| 1 | Preparation and planting | 0 | 10,437 | — |
| 2 | Maintenance | 0 | 5,346 | — |
| 3 | Maintenance | 0 | 5,346 | — |
| 4 | Partial production | 3,000 | 7,425 | 2.48 |
| 5 | Partial production | 6,000 | 7,425 | 1.24 |
| 6-20 | Full production | 10,000 | 9,394 | 0.94 |

TABLE B.2. Yields and costs for white currant production.

| Year | Stage of production | Yield (lb/acre) | Cost/acre ($) | Cost/pound ($) |
|------|---------------------|-----------------|---------------|----------------|
| 1 | Preparation and planting | 0 | 10,437 | — |
| 2 | Maintenance | 0 | 5,346 | — |
| 3 | Maintenance | 0 | 5,346 | — |
| 4 | Partial production | 4,000 | 7,425 | 1.86 |
| 5 | Partial production | 8,000 | 7,425 | 0.93 |
| 6-20 | Full production | 12,000 | 9,394 | 0.78 |

TABLE B.3. Yields and costs for black currant production.

| Year | Stage of production | Yield (lb/acre) | Cost/acre ($) | Cost/pound ($) |
|------|---------------------|-----------------|---------------|----------------|
| 1 | Preparation and planting | 0 | 10,437 | — |
| 2 | Maintenance | 0 | 5,346 | — |
| 3 | Maintenance | 0 | 5,346 | — |
| 4 | Partial production | 1,500 | 7,425 | 4.95 |
| 5 | Partial production | 3,000 | 7,425 | 2.48 |
| 6-20 | Full production | 5,500 | 9,394 | 1.70 |

TABLE B.4. Yields and costs for gooseberry production.

| Year | Stage of production | Yield (lb/acre) | Cost/acre ($) | Cost/pound ($) |
|------|---------------------|-----------------|---------------|----------------|
| 1 | Preparation and planting | 0 | 10,437 | — |
| 2 | Maintenance | 0 | 5,346 | — |
| 3 | Maintenance | 0 | 5,346 | — |
| 4 | Partial production | 1,500 | 7,425 | 4.95 |
| 5 | Partial production | 3,000 | 7,425 | 2.48 |
| 6 | Partial production | 4,500 | 7,425 | 1.65 |
| 7-20 | Full production | 6,000 | 9,394 | 1.57 |

TABLE B.5. Estimated equipment costs for a two-acre currant or gooseberry farm.

| Item | 2002 price ($) | Annual use | Years to trade | Cost/hour ($) | Cost/year ($) |
|---|---|---|---|---|---|
| Tractor (24 hp) | 19,220 | 50 hr | 20 | 19 | 961 |
| Trailer | 1,860 | 30 hr | 20 | 3 | 93 |
| Tractor-mounted sprayer | 992 | 10 hr | 20 | 5 | 50 |
| Cone fertilizer | 558 | 12 hr | 20 | 2 | 28 |
| Mower | 1,054 | 10 hr | 20 | 5 | 53 |
| Rotary tiller | 2,480 | 15 hr | 20 | 8 | 124 |
| Misc. equipment[a] | 775 | — | — | varies | 39 |

*Note:* Table assumes that all equipment is purchased new and used exclusively in the currant or gooseberry operation. The equipment is financed at 12 percent interest over 60 months with no down payment. Equipment that is needed only during preparation or otherwise infrequently, such as a seeder or roller, can be rented rather than purchased. Purchasing used equipment or using equipment for operations in addition to *Ribes* production will reduce the initial cost associated with the currant or gooseberry operation but can increase repair costs and down time.

[a]Backpack sprayer, shears, scale, buckets, 15 picking stands.

TABLE B.6. Typical permanently installed resources for a two-acre currant or gooseberry farm.

| Item | Size/type | 2002 purchase price ($) | Useful life (years) |
|---|---|---|---|
| Walk-in cooler | 10' × 16' | 13,600 | 20 |
| Herbivore fence | New Zealand | 860 | 20 |
| Irrigation system | Trickle | 4,620 | 20 |

*Note:* Table assumes that the cooler, fence, and irrigation system are purchased new and financed at 12 percent interest over 60 months with no down payment.

TABLE B.7. Trickle irrigation system for two acres of currants or gooseberries.

| Item | Size | Quantity | 2002 cost ($) |
|---|---|---|---|
| Mainline PVC | 1.5 inch | 400 feet | 280 |
| Risers PVC | 0.5 inch | 160 feet | 50 |
| Tubing | 0.5 inch | 7,200 feet | 320 |
| Fittings and tees | — | — | 185 |
| Emitters | 0.5 gal/hour | 3,600 | 580 |
| Valves | 1.5 inch | 3 | 185 |
| Filter (80 gal/hr) | — | 1 | 175 |
| Pump | 3 hp | 1 | 2,480 |
| Timer | — | 1 | 125 |
| Boxes for risers and valves | — | 39 | 240 |
| Total | | | 4,620 |

*Note:* The figures in this table do not include the cost of a well or access to surface water, nor do they include the cost of providing electrical power to the pump.

TABLE B.8. Total establishment costs (preparation and planting year) for two acres of currants or gooseberries.

| Activity | Machinery ($) | Labor ($) | Materials ($) | Service ($) | Total ($) | Your cost ($) |
|---|---|---|---|---|---|---|
| Soil tests | — | — | — | 100 | 100 | |
| Soil amendments | | | | | | |
| Lime[a] | | | | | | |
| Sulfur[a] | | | | | | |
| Gypsum[a] | | | | | | |
| Other[a] | | | | | | |
| | | | Variable costs | | | |
| Spray nonselective herbicide | 2 | 21 | 75 | — | 98 | |
| Rototill | 9 | 83 | — | — | 92 | |
| Pack ground | 2 | 18 | — | — | 20 | |
| Seed manure crop | 1 | 8 | 30 | — | 39 | |
| Spot spray weeds | — | 20 | 15 | — | 35 | |
| Custom backhoe | — | — | — | 840 | 840 | |
| Install trickle irrigation system | — | 660 | — | — | 660 | |
| Install herbivore fence | — | 462 | — | — | 462 | |
| Rototill manure crop | 9 | 83 | — | — | 92 | |
| Pack ground | 2 | 18 | — | — | 20 | |
| Lay out planting rows | — | 83 | — | — | 83 | |

TABLE B.8 (continued)

| Activity | Machinery ($) | Labor ($) | Materials ($) | Service ($) | Total ($) | Your cost ($) |
|---|---|---|---|---|---|---|
| | | | Variable costs | | | |
| Plant bushes | — | 1,340 | 7,533 | — | 8,873 | |
| Irrigate | 19 | 127 | — | — | 146 | |
| Mulch rows with sawdust | 2 | 306 | 1,080 | — | 1,388 | |
| Dormant fungicide spray | 2 | 21 | 100 | — | 123 | |
| Overhead (5 percent) | — | — | — | — | 654 | |
| Interest on operating capital (12 percent) | — | — | — | — | 1,647 | |
| Total variable costs | | | | | 15,372 | |
| | | | Fixed costs | | | |
| Machinery and equipment | — | — | — | — | 2,898 | |
| Permanent fixtures | — | — | — | — | 2,544 | |
| Property taxes | — | — | — | — | 60 | |
| Total fixed costs | | | | | 5,502 | |
| Total establishment costs | | | | | 20,874 | |

[a]These materials may or may not be needed, depending on your soil. Apply fertilizers and soil amendment materials in accordance with analytical lab recommendations.

TABLE B.9. Annual maintenance costs (years 2 and 3) for two acres of currants or gooseberries.

| Activity | Machinery ($) | Labor ($) | Materials ($) | Service ($) | Total ($) | Your cost ($) |
|---|---|---|---|---|---|---|
| | | Variable costs | | | | |
| Soil analysis | — | — | — | 35 | 35 | _____ |
| Spring oil/fungicide spray | 3 | 42 | 202 | — | 247 | _____ |
| Sulfur sprays (2) | 7 | 83 | 10 | — | 100 | _____ |
| Insecticide sprays (2) | 7 | 83 | 42 | — | 132 | _____ |
| Rototill alleys (year 2 only) | 9 | 83 | — | — | 92 | _____ |
| Pack ground (year 2 only) | 2 | 18 | — | — | 20 | _____ |
| Seed alley cover crop (year 2) | 1 | 5 | 177 | — | 183 | _____ |
| Pack ground (year 2 only) | 2 | 18 | — | — | 20 | _____ |
| Spot spray weeds | — | 20 | 15 | — | 35 | _____ |
| Hand weeding | — | 1,800 | — | — | 1,800 | _____ |
| Fertilize berries | — | 251 | 44 | — | 295 | _____ |
| Fertilize cover crop | 1 | 48 | 22 | — | 71 | _____ |
| Mow cover crop | 1 | 11 | — | — | 12 | _____ |
| Rodent control | — | 20 | 10 | — | 30 | _____ |
| Foliage analysis | — | — | — | 50 | 50 | _____ |
| Drip irrigation expenses | — | 150 | 200 | — | 350 | _____ |

TABLE B.9 *(continued)*

| Activity | Machinery ($) | Labor ($) | Materials ($) | Service ($) | Total ($) | Your cost ($) |
|---|---|---|---|---|---|---|
| | | | Variable costs | | | |
| Herbivore fence mainte-nance | — | 70 | 30 | — | 100 | |
| Fall dormant fungicide spray | 2 | 21 | 100 | — | 123 | |
| Overhead (5 percent) | — | — | — | — | 185 | |
| Interest on operating cap. (12 percent) | — | — | — | — | 466 | |
| Total variable costs | | | | | 4,346 | |
| | | | Fixed costs | | | |
| Establishment costs | — | — | — | — | 854 | |
| Machinery and equip-ment | — | — | — | — | 2,898 | |
| Permanent fixtures | — | — | — | — | 2,544 | |
| Property tax | — | — | — | — | 60 | |
| Total fixed costs | | | | | 6,346 | |
| Total costs | | | | | 10,692 | |

*Note:* Fixed costs for establishment (years 1 and 2), machinery, and permanent fixtures assume initial costs are amortized over the 20-year expected life of the stand. Figures represent principal and interest payments based on 12 percent interest.

TABLE B.10. Annual partial production costs (years 4 and 5 for currants and 4 to 6 for gooseberries) per two acres.

| Activity | Machinery ($) | Labor ($) | Materials ($) | Service ($) | Total ($) | Your cost ($) |
|---|---|---|---|---|---|---|
| | | Variable costs | | | | |
| Soil analysis | — | — | — | 35 | 35 | _____ |
| Pruning | — | 250 | — | — | 250 | _____ |
| Spring oil/fungicide spray | 3 | 42 | 202 | — | 247 | _____ |
| Sulfur sprays (2) | 7 | 83 | 10 | — | 100 | _____ |
| Insecticide sprays (2) | 7 | 83 | 42 | — | 132 | _____ |
| Spot spray weeds | — | 20 | 15 | — | 35 | _____ |
| Hand weeding | — | 1,800 | — | — | 1,800 | _____ |
| Fertilize berries | — | 251 | 44 | — | 295 | _____ |
| Fertilize cover crop | 1 | 48 | 22 | — | 71 | _____ |
| Bee hive rental | — | — | — | 100 | 100 | _____ |
| Rent portable toilet for pickers | — | — | — | 50 | 50 | _____ |
| Harvest | — | 2,850 | 500 | — | 3,350 | _____ |
| Operation of cooler | — | — | — | 100 | 100 | _____ |
| Mow cover crop | 1 | 11 | — | — | 12 | _____ |
| Rodent control | — | 20 | 10 | — | 30 | _____ |
| Foliage analysis | — | — | — | 50 | 50 | _____ |
| Drip irrigation expenses | — | 150 | 200 | — | 350 | _____ |

## TABLE B.10 (continued)

| Activity | Machinery ($) | Labor ($) | Materials ($) | Service ($) | Total ($) | Your cost ($) |
|---|---|---|---|---|---|---|
| | | | Variable costs | | | |
| Herbivore fence maintenance | — | 70 | 30 | — | 100 | |
| Fall dormant fungicide spray | 2 | 21 | 100 | — | 123 | |
| Overhead (5 percent) | — | — | — | — | 362 | |
| Interest on operating cap. (12 percent) | — | — | — | — | 911 | |
| Total variable costs | | | | | 8,503 | |
| | | | Fixed costs | | | |
| Establishment costs | — | — | — | — | 854 | |
| Machinery and equipment | — | — | — | — | 2,898 | |
| Permanent fixtures | — | — | — | — | 2,544 | |
| Property tax | — | — | — | — | 60 | |
| Total fixed costs | | | | | 6,346 | |
| Total costs | | | | | 14,849 | |

Note: Fixed costs for establishment (years 1 and 2), machinery, and permanent fixtures assume initial costs are amortized over the 20-year expected life of the stand. Figures represent principal and interest payments based on 12 percent interest.

TABLE B.11. Annual full production costs (years 6 to 20 for currants and 7 to 20 for gooseberries) per two acres.

| Activity | Machinery ($) | Labor ($) | Materials ($) | Service ($) | Total ($) | Your cost ($) |
|---|---|---|---|---|---|---|
| | | | Variable costs | | | |
| Soil analysis | — | — | — | 35 | 35 | _____ |
| Pruning | — | 250 | — | — | 250 | _____ |
| Spring oil/fungicide spray | 3 | 42 | 202 | — | 247 | _____ |
| Sulfur sprays (2) | 7 | 83 | 10 | — | 100 | _____ |
| Insecticide sprays (2) | 7 | 83 | 42 | — | 132 | _____ |
| Spot spray weeds | — | 20 | 15 | — | 35 | _____ |
| Hand weeding | — | 1,800 | — | — | 1,800 | _____ |
| Fertilize berries | — | 251 | 44 | — | 295 | _____ |
| Fertilize cover crop | 1 | 48 | 22 | — | 71 | _____ |
| Bee hive rental | — | — | — | 100 | 100 | _____ |
| Rent portable toilet for pickers | — | — | — | 50 | 50 | _____ |
| Harvest | — | 5,700 | 1,000 | — | 6,700 | _____ |
| Operation of cooler | — | — | — | 100 | 100 | _____ |
| Mow cover crop | 1 | 11 | — | — | 12 | _____ |
| Rodent control | — | 20 | 10 | — | 30 | _____ |
| Foliage analysis | — | — | — | 50 | 50 | _____ |
| Drip irrigation expenses | — | 150 | 200 | — | 350 | _____ |
| Herbivore fence maintenance | — | 70 | 30 | — | 100 | _____ |

245

TABLE B.11 *(continued)*

| Activity | Machinery ($) | Labor ($) | Materials ($) | Service ($) | Total ($) | Your cost ($) |
|---|---|---|---|---|---|---|
| | | | Variable costs | | | |
| Fall dormant fungicide spray | 2 | 21 | 100 | — | 123 | ___ |
| Overhead (5 percent) | — | — | — | — | 529 | ___ |
| Interest on operating cap. (12 percent) | — | — | — | — | 1,333 | ___ |
| Total variable costs | | | | | 12,442 | ___ |
| | | | Fixed costs | | | |
| Establishment costs | — | — | — | — | 854 | ___ |
| Machinery and equipment | — | — | — | — | 2,898 | ___ |
| Permanent fixtures | — | — | — | — | 2,544 | ___ |
| Property tax | — | — | — | — | 60 | ___ |
| Total fixed costs | | | | | 6,346 | ___ |
| Total costs | | | | | 18,788 | ___ |

*Note:* Fixed costs for establishment (years 1 and 2), machinery, and permanent fixtures assume initial costs are amortized over the 20-year expected life of the stand. Figures represent principal and interest payments based on 12 percent interest.

# References

Aaltonen, M. and P. Dalman. (1993). The effect of fertilization on leaf and soil analyses of *Ribes rubrum* L. and *Ribes nigrum* L. *Acta Hort.* 352:21-28.

Ammonius. (1539). Med. Hort: 310. Cited in Hedrick (1925).

Amrine, J.W. (1992). Eriophyid mites on *Ribes*. In K.E. Hummer (Ed.), *Proceedings for the* Ribes *risk assessment workshop* (pp. 17-20, 45-75). Corvallis, OR: U.S. Department of Agriculture, Agricultural Research Service, National Clonal Germplasm Repository.

Anderson, H.W. (1956). *Diseases of fruit crops.* New York: McGraw-Hill.

Anderson, M.M. (1977). Breeding black currant for northern regions of the U.K.: Juice quality. In *1976 Report of the Scottish Horticultural Research Institute* (pp. 48-49), Invergowrie, Dundee, Scotland.

Andersson, J. and E. von Sydow. (1964). The aroma of black currant, I: Higher boiling compounds. *Acta. Chem. Scand.* 18:1105-1114.

Andersson, J. and E. von Sydow. (1966). The aroma of black currants, II: Lower boiling compounds. *Acta. Chem. Scand.* 20:522-528.

Anonymous. (2000). Nova: A new powdery mildew fungicide. *SC Pumpkin News* 5(1):1-2.

Arasu, N.T. (1968). Overcoming self-incompatibility by irradiation. In *1967 Report of the East Malling Research Station* (pp. 109-112), East Malling, UK.

Auchter, E.C. and H.B. Knapp. (1937). *Orchard and small fruit culture.* New York: John Wiley and Sons.

Audette, M. and M.J. Lareau. (1996). *Currants and gooseberries culture guide.* Québec City, Canada: Conseils des productions végétales du Québec Inc.

Badescu, G. and L. Badescu. (1976). New aspects of black currant cultivation. *Productia Vegetala, Horticultura* 25:23-27.

Bailey, L.H. (1897). *The principles of fruit-growing.* New York: The MacMillan Company.

Baldini, E. and P.L. Pisani. (1961). Research on the biology of flowering and fruiting in blackcurrants [in Italian]. *Riv. Ortoflorofruttic. Ital.* 45:619-639.

Barney, D. and D. Gerton. (1992). An evaluation of the *Ribes* collection at the National Germplasm Repository. Proceedings of the International *Ribes* Association Conference, Minneapolis, MN, January.

Barney, D., R. Walser, S. Nelson, C. Williams, and V. Jolley. (1984). Control of iron chlorosis in apple trees with injections of ferrous sulfate and ferric citrate and with soil-applied iron-sul. *J. Plant Nutr.* 7:313-317.

Batzer, U. and H.U. Helm. (1999). Lagerung von Beerenobst [Storage of small fruits]. *Erwerbsobstbau* 41:51-55.

Bauer, R. (1955). Resistance problems in the genus *Ribes* and possibilities of their solution by intra- and intersectional crosses [in German, English summary]. Report of the 14th International Horticultural Congress (pp. 685-696), Sheveningen, the Netherlands.

Berger, A. (1924). *A taxonomic review of currants and gooseberries.* Bulletin 109. Ithaca: New York State Agricultural Experiment Station.

Bould, C. (1969). Leaf analysis as a guide to the nutrition of fruit crops, VIII: Sand culture N, P, K, Mg experiments with black currant (*Ribes nigrum* L.). *J. Sci. Food Agric.* 20:172-181.

Bradfield, E.G. (1969). The effect of intensity of nutrient supply on growth, yield and leaf composition of black currant grown in sand culture. *J. Hort. Sci.* 44:211-218.

Brennan, R.M. (1990). Currants and gooseberries *(Ribes).* In J. Moore and J. Ballington (Eds.), *Genetic resources of temperate fruit and nut crops* (pp. 459-488). Wageningen, the Netherlands: International Society for Horticultural Science.

Brennan, R.M. (1996). Currants and gooseberries. In J. Janick and J. N. Moore (Eds.), *Fruit breeding,* Volume 2, *Vine and small fruit crops* (pp. 191-295). New York: John Wiley and Sons.

Byers, R., D. Carbaugh, and C. Presley. (1990). Screening of odor and taste repellants for control of white-tailed deer browse to apples or apple shoots. *J. Env. Hort.* 8:185-189.

California Rare Fruit Growers, Inc. (2002). Gooseberry fruit facts. Available at <http://www.crfg.org/pubs/ff/gooseberry.html>.

Camerarius. (1586). Epistle. pp. 88. Cited in Hedrick (1925).

Camerarius. (1587). Hort. 141. Cited in Hedrick (1925).

Card, F.W. (1907). *Bush fruits.* London: The Macmillan Co.

Carlson, C.E. (1978). *Noneffectiveness of* Ribes *eradication as a control of white pine blister rust in Yellowstone National Park.* Report No. 78-18. Missoula, MT: Forest Insect and Disease Management, USDA Forest Service, Northern Region.

Chrapkowska, K. and K. Rogalinski. (1975). The effect of some meteorological conditions on vitamin C content in ten blackcurrant varieties [in Polish]. *Zeszyty Naukowe Akademii Rolniczej w Krakowie, Lesnictwo* 8:65-76.

Costin, J.J. and T.A. Kenny. (1972). Effects of 2-chloroethylyphosphonic acid (Ethrel) on gooseberries. *Pestic. Sci.* 3:545-550.

Coville, F.V. and N.L. Britton. (1908). Grossulariaceae. *N. Amer. Fl.* 22:193-225.

Cronquist, A. (1981). *An integrated system for classification of flowering plants.* New York: Columbia University Press.

Culverwell, W. (1883). Letters. *Gardeners Chronicles* 19:635.

Cuvasina, N.P. (1961). The effect of gibberellin on crossability between distantly related plants [in Russian]. *Tr. Cent. Genet. Lab. Micurina* 7:183-189.

Cuvasina, N.P. (1962). The effect of giberellic acid on the crossability of different species in the genus *Ribes* L. [in Russian]. Report of the Soviet Scientists (pp. 123-128), 16th International Horticulture Congress, Moscow.

Dale, A. (1992). Black currant potential in North America. In K.E. Hummer (Ed.), *Proceedings for the* Ribes *risk assessment workshop* (pp. 23-26). Corvallis, OR: U.S. Department of Agriculture, Agricultural Research Service, National Clonal Germplasm Repository.

Dale, A. (2000). Black plastic mulch and between-row cultivation increase black currant yields. *HortTechnology* 10:307-308.

Dale, A. and K. Schooley. (1999). *Currants and gooseberries*. Fact Sheet 98-095. Toronto: Ontario Ministry of Agriculture and Food.

Dalechamp, J. (1587). Hist. Gen. (usually referred to as Lugd.) 1:131. Cited in Hedrick (1925).

Darrow, G.M. (1919). *Currants and gooseberries*. USDA Farmers Bulletin 1024. Washington, DC: U.S. Government Printing Office.

Darrow, G.M. (1946). *Currants and gooseberries: Their culture and relation to white pine blister rust*. USDA Farmers Bulletin No. 1398. Washington, DC: U.S. Government Printing Office.

de Janczewski, E. (1907). Monograph of the currants *Ribes* L. [in French]. *Mem. Soc. Phys. Hist. Nat. Geneve* 35:199-517.

Dennis, C. (1983). Soft fruits. In C. Dennis (Ed.), *Postharvest pathology of fruits and vegetables* (pp. 23-42). New York: Academic Press.

Dow, A.I. (1980). *Critical nutrient ranges in northwest crops*. Western Regional Extension Publication (WREP) 43. Pullman: Washington State University.

Downing, A.J. (1845). *The fruits and fruit trees of America*. New York: Wiley and Putnam.

Duka, S. Kh. (1940). A new form of berry: Black currant x gooseberry. *Jarovizacija* 3(30):119-122.

Elsakova, S.D. (1972). The biological characteristics of black currant under the conditions of Murmansk province [in Russian]. In *Kul'tura chern. smorodiny* (pp. 58-62). Moscow.

Engler, A. and K. Prantl. (1891). Ribesioideae. *Naturl. Pflanzenfam.* 3:97-142.

Eppler, A. (1989). Growing gooseberries and currants as single vertical cordons: An alternative to bush grown plants. *The* Ribes *Reporter* 1(2):3, 5.

Felski, J. and A. Brzezinska. (1988). Levels of input and efficiency of direct labor with various methods of harvesting of black-currants on individual family farms in Poland. *Acta Hortic* 223:145-148.

Fernqvist, I. (1961). Investigations on floral biology in blackcurrants, redcurrants, and gooseberries [in Swedish]. *Kgl. LantbrAkad. Tidsk.* 100:357-397.

Food and Agriculture Organization (FAO). (2002). Statistical database. Available at <http://apps.fao.org/>.

Free, J.B. (1968). The pollination of blackcurrants. *J. Hort. Sci.* 43:69-73.

Fuchs (Fuchsius), L. (1542). Di Historia Stirpium.

Gerarde, J. (1597). *General Historie of Plants.* London: John Norton.

Gorskov, I.S. (1940). From the work of the Central Breeding and Genetics Laboratory of I.V. Michurin. *Jarovizacija* 3(30):152-156.

Gubenko, A.P., V.S. Il'in, and M.A. Chirkova. (1976). Breeding black currant and gooseberry for high content of ascorbic acid [in Russian]. In *Biol. Aktiv. Veschevstva plodov I yagod* (pp. 69-71). Moscow.

Hansen, E.M. (1979). Pathogenic variation in *Cronartium ribicola* pathogenicity to *Ribes* and *Pinus monticola.* In *The Station 1979* (pp. 59-66). Fort Collins, CO: U.S. Forest Service Rocky Mountain Forest Range Experiment Station.

Harborne, J.B. and E. Hall. (1964). Plant polyphenols, 13: The systematic distribution and origin of anthocyanins containing branched trisaccharides. *Phytochemistry* 3:453-463.

Hardenburg, R.E., A.E. Watada, and C.Y Wang. (1986). *The commercial storage of fruits, vegetables, and florist and nursery stocks.* USDA Agriculture Handbook No. 66 (revised). Washington, DC: U.S. Government Printing Office.

Harmat, L., A. Porpaczy, D.G. Himelrick, and G.J. Galletta. (1990). Currant and gooseberry management. In G. Galletta and D. Himelrick (Eds.), *Small fruit crop management* (pp. 245-272). Englewood Cliffs, NJ: Prentice Hall.

Hedrick, U.P. (1925). *The small fruits of New York.* Report of the New York State Agricultural Experiment Station for the year ending June 30, 1925. Albany, NY: J.B. Lyon Co.

Heiberg, N. (1986). *Bud dormancy and root development in cuttings of black currant* (Ribes nigrum *L.*). Report No. 118. As, Norway: Agricultural University of Norway, Department of Pomology.

Hofman, K. (1963). Fruit set in a number of blackcurrants [in Dutch]. *Fruiteelt* 53:334-335.

Holubowicz, T. and K. Bojar. (1982). Cold tolerance studies of one year shoots of eight black currant cultivars. *Fruit Sci. Rpt.* 9:91-99.

Hooper, F.C. and A.D. Ayers. (1950). The enzymatic degradation of ascorbic acid by substances occurring in black currants. *J. Sci. Food Agric.* 1:5-8.

Hughes, H.M. (1972). Soil and bush management studies on gooseberries. *Exp. Hort.* 24:43-49.

Hukkanen, A.T., T.P. Mikkonen, K.R. Maatta, A.R. Torronen, S.O. Karenlampi, H.I. Kokko, and R.O. Karjalainen. (1993). Variation in flavonol content among blackcurrant cultivars. *Acta Hort.* 352:121-124.

Hummer, K. E. and D. Picton. (2001). Oil application reduces powdery mildew severity in red and black currants. *HortTechnology* 11(3):445-446.

Hunter, A.W. (1950). Small fruits: Black currants. In *1934-1938 Progress Report of the Central Experiment Farm* (pp. 26-29), Ottawa, Ontario, Canada.

Hunter, A.W. (1955). Black currants. In *1949-1953 Progress report of the Central Experiment Farm* (pp. 28-29), Ottawa, Ontario, Canada.

Jennings, D.L., M.M. Anderson, and R.M. Brennan. (1987). Raspberry and blackcurrant breeding. In A.J. Abbot and R.K. Atkin (Eds.), *Improving vegetatively propagated crops* (pp. 135-147). London: Academic Press.

Jones, A.T. (1992). Defining the problem: Reversion disease and eriophyid mite vectors in Europe. In K.E. Hummer (Ed.), *Proceedings for the* Ribes *risk assessment workshop* (pp. 1-4). Corvallis, OR: U.S. Department of Agriculture, Agricultural Research Service, National Clonal Germplasm Repository.

Kader, A.A. (1992). Postharvest biology and technology: An overview. In A.A. Kader (Ed.), *Postharvest technology of horticultural crops* (pp. 15-20). Publication No. 3311. Oakland: University of California Division of Agriculture and Natural Resources.

Karnatz, A. (1969). Raising *Ribes nigrum* seedlings under cover [in German]. *Mitt. Klosterneuburg* 19:319-321.

Kasmire, R.F. and J.F. Thompson. (1992). Selecting a cooling method. In A.A. Kader (Ed.), *Postharvest technology of horticultural crops* (pp. 63-68). Publication No. 3311. Oakland: University of California Division of Agriculture and Natural Resources.

Keeble, F. and A.N. Rawes. (1948). *Hardy fruit growing.* London: MacMillan and Co.

Keep, E. (1962). Interspecific hybridization in *Ribes. Genetica* 33:1-23.

Keep, E. (1975). Currants and gooseberries. In J. Janick and J. Moore (Eds.), *Advances in fruit breeding* (pp. 197-268). West Lafayette, IN: Purdue University Press.

Keep, E. (1976). Currants: *Ribes* spp. (Grossulariaceae). In N.W. Simmonds (Ed.), *Evolution of crop plants* (pp. 145-148). London: Longman.

Keep, E. (1977). North European cultivars as donors of resistance to American powdery mildew in black currant breeding. *Euphytica* 26:817-823.

Keep. E. (1985). The mildew resistance gene $Sph_2$ in relation to gall mite resistance in the black currant. *Euphytica* 34:865-868.

Keep, E. (1986). Cytoplasmic male sterility, resistance to gall mite and mildew, and season of leafing out in black currants. *Euphytica* 35:843-855.

Keep, E., J.H. Parker, and V.H. Knight. (1977). Black currant breeding using related species as donors. *ARC Research Review* 3:86-88.

Kennedy, C.T. (1990). And now jostaberry. *Fruit Gardener* 22(3):12-13, 22.

Klambt, H.-D. (1958). Studies on pollination behavior in black- and red currants [in German]. *Gartenbauwissenschaft.* 23:9-28.

Knight, R.L. and E. Keep. (1957). Fertile black currant-gooseberry hybrids. In *1957 Report of the Malling Research Station* (pp. 73-74), Malling, UK.

Knight, V.H. (1983). The effect of donor species used in breeding on juice quality of black currants (*Ribes nigrum* L.). *J. Hort. Sci.* 58:63-71.

Komarov, V.L. (Ed.) (1971). *Flora of the USSR,* Volume 9. In *Ribesioidea* Engl. (pp. 175-208) [translated from Russian into English by the Israel Program for Scientific Translations, Jerusalem]. London: Keter.

Kronenberg, H.G. (1964). Some varietal differences in redcurrant. *Hort. Res.* 3:72-78.

Kronenberg, H.G. and K. Hofman. (1965). Research on some characters in blackcurrant progenies. *Euphytica* 14:23-35.

Kuminov, E.P. (1962). Self-fertility and cross-fertility of Siberian currant varieties [in Russian]. *Sel'sk Hosjajstv. Sibir.* 12:57-59.

Lamarck, J.B. and A.P. De Candolle. (1805). *Flore francaise.* Paris: Desray.

Lantin, B. (1970). Importance of cross-pollination in the black currant [in French]. *Pomol. Fr.* 8:237-243.

Larsson, L., B. Stenberg, and L. Torstensson. (1997). Effects of mulching and cover cropping on soil microbial parameters in the organic growing of black currant. *Comm. Soil Sci. Plant Anal.* 28:913-925.

Latrasse, A. and B. Lantin. (1974). Varietal differences between the monoterpene hydrocarbons of the essential oil of black currant buds [in French]. *Ann. Techn. Agric.* 23:65-74.

Ledeboer, M. and I. Rietsema. (1940). Unfruitfullness in black currants. *J. Pomol.* 18:177-181.

LeLous, J., B. Majoie, J.-L. Moriniere, and E. Wulfert. (1975). Studies of the flavonoids of *Ribes nigrum* [in French]. *Ann. Pharm. Fr.* 33:393-399.

Lemmetty, A., S. Latvala, A.T. Jones, W.J. McGavin, P. Susi, and K. Lehto. (1997). Properties and affinities with nepoviruses of blackcurrant reversion associated virus, a new virus isolated from blackcurrant (*Ribes nigrum* L.) affected with the severe form of reversion disease. *Phytopathology* 87: 404-413.

Lenartowicz, W., W. Polcharski, and L. Wlodek. (1976). The influence of fertilisation on the quality of small fruits, I: The influence of mineral fertilization of black currant on the chemical composition of extracted juice. *Fruit Sci. Rep.* 3:43-50.

Lewis, M.J., H.V. May, and A.A. Williams. (1980). Fruit quality: Black currant. In *1978 Report of the Long Ashton Research Station* (pp. 156-157), Bristol, UK.

Linnaeus, C. (1737). Hortus cliffortianus. Amsterdam.

Linnaeus, C. (1738). Genera plantarum.

Lobel. (1576). Obs: 615. Cited in Hedrick (1925).

Lobel. (1591). Obs: 2002. Cited in Hedrick (1925).

Lucka, M., W. Lech, and K. Dobrinska. (1972). The course of the pollination process and research on the degree of self pollination in the black currant [in German]. *Zestyly Nauk. Wyzsej. Szkoly. Roln. Krakow* 2:89-102.

Luckwill, L.C. (1948). A note on the unfruitfullness of a rogue strain of the black currant variety 'Invincible Giant Prolific'. In *1978 Report of the Long Ashton Research Station* (pp. 22-25), Bristol, UK.

Mage, F. (1976). *Bud dormancy and root formation on cuttings of currants.* Report no. 89. As, Norway: Agricultural University of Norway, Department of Pomology.

Mapson, L.W. (1970). Vitamins in fruits. In A.C. Hulme (Ed.), *The biochemistry of fruits and their products,* Volume 1 (pp. 369-384). London: Academic Press.

Marriott, R. (1988). Isolation and analysis of blackcurrant (*Ribes nigrum* L.) leaf oil. In B.M. Lawrence, B.D. Mookherjee, and B.J. Willis (Eds.), *Flavors and fragrances: A world perspective.* Amsterdam: Elsevier Science Publishers.

Massachusetts Company. (1629). Massachusetts Records V. 1:24. Boston: Commonwealth of Massachusetts.

Matthiolus. (1558). Comment: 101. Cited in Hedrick (1925).

Melekhina, A.A. (1968). Effect of gamma irradiation on the viability of pollen in intraspecific crosses of black currants. In *Crop plants in the national economy* 4 [in Russian]. Zinatne, Riga 4 (from Referat Zh. 1970:145-155). Riga, Latvia: Zinatne Press.

Melekhina, A.A., B.B. Yankelevich, and M.A. Eglite. (1980). Prospects for the use of *Ribes petiolare* Dougl. in breeding blackcurrant [in Russian]. *Latvijas PSR Zinatu Akad.* 3:116-123.

Mizaldus. (1560). Secretorum: 105. Cited in Hedrick (1925).

Morgan, J.M., M.D. Morgan, and J.H. Wiersma. (1980). *Introduction to environmental science.* San Francisco: W.H. Freeman Co.

Moyer, R.A., K.E. Hummer, C.E. Finn, B. Frei, and R.E. Wrolstad. (2002). Anthocyanins, phenolics, and antioxidant capacity in diverse small fruits: *Vaccinium, Rubus,* and *Ribes. J. Agric. Food Chem.* 50:519-525.

Moyer, R., K. Hummer, R. Wrolstad, and C. Finn. (2002). Antioxidant compounds in diverse *Ribes* and *Rubus* germplasm. *Acta Hort.* 585:501-505.

Nemethy, L. (1977). *Technical testing of shaker type black currant harvesters* [Vibracios feketeribiszke betakarito gepek muszaki vizsgalata] [in Hungarian with English summary]. Godollo: Mezogazdasagi Gepkiserleti Intezet.

Nes, A. (1976). Cross pollination in black currants [in Norwegian]. *Forskning og Forsok i Landbruket* 27:717-730.

Neumann, U. (1955). The importance of fertilization conditions and cultural measures for premature fruit drop in black currants [in German]. *Arch. Gartenb.* 3:339-354.

Nilsson, F. (1969). Ascorbic acid in blackcurrants. *Lantbrukshogskolans Annlr* 35:43-59.

Nilsson, F. and V. Trajkovski. (1977). Colour pigments in species and hybrids of the genus *Ribes* [in Swedish]. *Lantbrukshogskolans Meddelanden* 282:1-20.

Nursten, H.E. and A.A. Williams. (1967). Fruit aromas: A survey of components identified. *Chem. Ind.* 12:486-497.

Olander, S.A. (1993). High density cultural system for black currants (*Ribes nigrum* L.). *Acta Hort.* 352:71-78.

Osipov, K.V. (1968). Regular annual yields of currants [in Russian]. *Sadovodstvo* 8:41.

Oydvin, J. (1973). Inheritance of fruit color in red currant [in Norwegian]. *Forskning Forsok Landbruk.* 24:539-542.

Oydvin, J. (1978). Description of three new red currant cultivars 'Fortun', 'Nortun', and 'Jontun', and characteristics of parents [in Norwegian]. *Gartnerysket* 68: 452-454, 456-457.

Parkinson, J. (1629). *Paradisi in sole paradisus terrestris.* London: Humphrey, Turner, and Robert Young.

Phillips, H. (1820). Pomarium Britannicum: 138. London.

Picton, D.D. and K.E. Hummer. (2003). Oil application reduces white pine blister rust severity in black currants. *Small Fruits Review* 2(1):43-48.

Pinaeus. (1561). Hist: 67. Cited in Hedrick (1925).

Potapenko, A.A. (1966). The choice of parental pairs in breeding self-fertile varieties of blackcurrants [in Russian]. *Nauk. Tr. Omsk. Sel'skohoz. Inst.* 64:104-108.

Prange, R.K. (2002). Currant, gooseberry and elderberry. In K.C. Gross, C.Y. Wang, and M. Saltveit (Eds.), *The commercial storage of fruits, vegetables, and florist and nursery stocks.* U.S. Department of Agriculture Handbook Number 66. Available at <http://www.ba.ars.usda.gov/hb66/contents.html>.

Pritts, M., J. Hancock, B. Strik, M. Eames-Sheavly, W. Autio, M. Burgett, R. Childs, J. Clark, R. Cook, A. Draper, et al. (1992). *Highbush blueberry production guide.* NRAES 55. Ithaca, NY: Northeast Regional Agricultural Engineering Service.

Pritts, M. and W. Wilcox. (1986). *1986 Cornell recommendations for small fruit production.* Ithaca, NY: Cornell University.

Pscheidt, J.W. (2000). Gooseberry and currant—Powdery mildew. In *An online guide to plant disease control,* Oregon State University Extension. Available at <http://plant-disease.ippc.orst.edu/disease.cfm?RecordID=512>.

Pscheidt, J.W. and C.M. Ocamb (Eds.). (2001). *Pacific Northwest 2001 plant disease management handbook.* Corvallis: Oregon State University.

Raincikova, G.P. (1967). Germination of blackcurrant pollen and conditions influencing its viability. In *Fruit and berry crops* (pp. 232-237) [in Russian]. Minsk: Urozaj.

Rake, B.A. (1958). The history of gooseberries in England. *Fruit Yearb.* 10:84-87.

Rea, J. (1665). *Flora, ceres and pomona.* London: Richard Marriott.

Rehder, A. (1954). *Manual of cultivated trees and shrubs.* New York: Macmillan.

Rehder, A. (1986). *Manual of cultivated trees and shrubs,* Revised second edition. Portland, OR: Discorides Press.

Robinson, J.E., K.M. Browne, and W.G. Burton. (1975). Storage characteristics of some vegetables and soft fruits. *Ann. Appl. Biol.* 81:339-408.

Roelofs, F.P.M.M. and A.J.P.v. Waart. (1993). Long-term storage of red currants under controlled atmosphere conditions. *Acta Hort.* 352:217-222.

Rudowski, R.M. (2002). *Still in our memories: Currants and gooseberries.* Boonville, CA: The International *Ribes* Association.

Ruellius, J. (1536). De natura stirpium libri tress: Paris. Cited in Hedrick (1925).

Ryall, A.L. and W.T. Pentzer. (1982). *Handling, transportation and storage of fruits and vegetables,* Volume 2, *Fruits and tree nuts,* Second edition. Westport, CT: AVI.

Salamon, Z. (1993). Mechanical harvest of black currants and their sensitivity to damage. *Acta Hort.* 352:109-112.

Salamon, Z. and D. Chlebowska. (1993). Preliminary trials with mechanical harvest of gooseberry. *Acta Hort.* 352:105-108.

Samorodova-Bianki, G.B., E.V. Volodina, and L.E. Baskokova. (1976). Chemical composition of the fruits of blackcurrants belonging to different groups of species [in Russian]. *Byul. Vses. Ord. Lenin Ord. Drzh. Narod. Inst. Rast. I.V. Vavilova* 61:61-63.

Schuricht, R. and D. Schwope. (1969). Yield and labor input with the cultivation of red currants on wire frames [Ertrage und aufwand beim anbau von rotten johannisbeeren an drahtgerusten] [in German with English summary]. *Arch. Gartenbau* 17(7):487-497.

Sears, M.F. (1925). *Productive small fruit culture*. Philadelphia: J.B. Lippencott Company.

Seljahudin, A. and A. Brozik. (1967). Fertilization conditions of berry fruit varieties, 3: Raspberry, black, red currant. *Acta Agron. Hung.* 15:187-198.

Serengovyj, P.Z. (1969). Blister rust of currant [in Hungarian]. *Z. Rast. Vredit.* 14:40-41.

Sergeeva, N.V. (1979). Self-fertility of gooseberry in relation to its genetic origin [in Russian]. *Sb. Nauk. Tr. VNII Sadavod. Im. I. V. Michurina* 29:36-38.

Sergeeva, N.V. (1980). Self-fertility in gooseberry [in Russian]. *Sadovodstvo* 7:27.

Shoemaker, J.S. (1948). *Small-fruit culture*. Toronto: McGraw Hill Book Company.

Sinnott, Q.P. (1985). A revision of *Ribes* L. subg. *Grossularia* (Mill.) pers. sect. *Grossularia* (Mill.) Nutt. (Grossulariaceae) in North America. *Rhodora* 87:189-286.

Smith, W.H. (1967). The storage of gooseberries. In *Ditton and Covent Garden Labs Annual Report, 1965-66* (pp. 13-14), Agriculture Research Council, UK.

Somorowski, K. (1964). Preliminary results on the breeding of blackcurrant [in Polish]. *Pr. Inst. Sadow. Skierniew.* 8:3-19.

Spayd, S.E., J.R. Morris, W.E. Ballinger, and D.G. Himelrick. (1990). Maturity standards, harvesting, postharvest handling, and storage. In G.J. Galletta and D.G. Himelrick (Eds.), *Small fruit crop management* (pp. 505-531). Englewood Cliffs, NJ: Prentice Hall.

Spinks, G.T. (1947). Black currant breeding at Long Ashton. In *Report of the Long Ashton Research Station* (pp. 35-43), Bristol, UK.

Spongberg, S.A. (1972). The genera of Saxifragaceae in the southeastern United States. *J. Arnold Arbor.* 53:409-498.

Stern, W.L., E.M. Sweitzer, and R.E. Phipps. (1970). Comparative anatomy and systematics of woody Saxifragaceae. *Ribes. J. Linn. Soc. Bot.* 63:215-237.

Story, A. and D.H. Simons. (1989). *Handling and storage practices for fresh fruit and vegetables*. Victoria: Australian United Fresh Fruit and Vegetable Assoc. Ltd.

Strik, B.C. and A.D. Bratsch. (1990). *Growing currants and gooseberries in your home garden*. EC 1361. Corvallis: Oregon State University.

Sturtevant, E.L. (1887). *Western New York Horticultural Society Proceedings*, p. 56. Cited in Hedrick (1925).

Szklanowska, K. and B. Dabska. (1993). The influence of insect pollinating on fruit setting of three black currant cultivars of (*Ribes nigrum* L.). *Acta Hort.* 352:223-230.

Tahvonen, R. (1979). Injury to currants during mechanical harvesting and subsequent fungal infection. *J. Sci. Agric. Soc. Finland* 51:421-431.

Tamas, P. (1964). Bush fruits [in Swedish]. In *1963 Report of the Balsgard Fruit Breeding Institute* (pp. 25-28), Fjalkestad, Sweden.

Tamas, P. (1968). Bush fruits [in Swedish]. In *1967 Report of the Balsgard Fruit Breeding Institute* (pp. 21-26), Fjalkestad, Sweden.

Tamas, P. and A. Porpaczy. (1967). Some physiological and breeding problems in the fertilization of the genus *Ribes,* 1: Variability in the compatibility in blackcurrants [in German]. *Zuchter* 37:232-238.

Taylor, J. (1989). Color stability of blackcurrant *(Ribes nigrum)* juice. *J. Sci. Food Agric.* 49:487-491.

Thomas, J.T. and W.H.S. Wood. (1909). *The American fruit culturist.* New York: Orange Judd Company.

Thompson, A.K. (1998). *Controlled atmosphere storage of fruits and vegetables.* Wallingford, UK: CAB International.

Tinklin, I.G. and W.W. Schwabe. (1970). Lateral bud dormancy in the blackcurrant *Ribes nigrum* L. *Ann. Bot.* 34:691-706.

Tinklin, I.G., E.H. Wilkinson, and W.W. Schwabe. (1970). Factors affecting flower initiation in the black currant *Ribes nigrum* L. *J. Hort. Sci.* 45:275-282.

Tolmacev, I.A. (1940). An experiment on overcoming incompatibility [in Hungarian]. *Jarovizacija* 5(32):125-126.

Tragus. (1552). de Stirpium: 994. Cited in Hedrick (1925).

Trajkovski, V. and R. Paasuke. (1976). Resistance to *Sphaerotheca mors-uvae* (Schw.) Berk. in *Ribes nigrum* L., 5: Studies on breeding black currants for resistance to *Sphaerotheca mors-uvae* (Schw.) Berk. *Swedish J. Agric. Res.* 6:201-214.

United States Department of Agriculture (USDA). (2003a). Federal and state plant quarantine summaries. Animal and Plant Health Inspection Service. Available at <http://www.aphis.usda.gov/npb/F&SQS/sqs.html>.

United States Department of Agriculture (USDA). (2003b). National Clonal Germplasm Repository, Corvallis, Oregon. Available at <http://www.ars-grin.gov/ars/PacWest/Corvallis/ncgr/>.

Utkov, Y.A. and P.I. Pilenko. (1984). Increasing the productivity of a berry-picking combine by increasing the rigidity of the àctivator picking pins. *Sov. Agric. Sci.* 5:78-81.

Van Meter, R.A. (1928). *Bush fruit production.* New York: Orange Judd Publishing Co.

Vang-Petersen, O. (1973). Leaf analysis, I: Nutrient contents in dry leaf matter in apple, pear, plum, cherry, black currant, and red currant in relation to the level of nitrogen, potassium, and magnesium. *Tidskr. Plant.* 77:393-398.

Voluznev, A.G. (1948). On the self-fertility of blackcurrants [in Russian]. *Sad Ogorod* 8:33-34.

Voluznev, A.G. (1968). Utilization of the Siberian species *Ribes nigrum* spp. *sibiricum* Pav. and *R. dikuscha* Fisch. in developing highly self-fertile varieties of blackcurrant in Byelorussia [in Russian]. In *Papers of a scientific conference on problems of genetics, breeding, and seed production in plants: Section top and small fruits and ornamental crops* (pp. 88-93). Gorki.

Voluznev, A.G. and G.P. Raincikova. (1974). The dependence of yield on the nature of pollination in blackcurrant varieties [in Russian]. *Plodovodstvo* 2:22-24.

Voluznev, A.G. and N.A. Zazulina. (1980). Selection of blackcurrant hybrids with a high content of vitamin C [in Russian]. *Plodovodstvo* 4:60-64.

Voscilko, M.E. (1969). *Some biological characteristics and morphological features of selected forms of wild berry fruits from the Salair ridge* [in Russian]. Novosibirsk: Nauka.

Watt, B.K. and A.L. Merrill. (1963). *Composition of foods*. USDA Agriculture Handbook 8. Washington, DC: U.S. Government Printing Office.

Webb, R.A. (1976). The components of yield in black currants. *Sci. Hort.* 4:247-254.

Wellington, R., R.G. Hatton, and J. Amos. (1921). The "running off" of blackcurrants. *J. Pomol.* 2:160-198.

Westwood, M.N. (1978). *Temperate-zone pomology*. San Francisco: W.H. Freeman and Co.

William, R., D. Ball, T. Miller, R. Parker, J. Yenish, T. Miller, D. Morishita, and P. Hutchinson. (2001). *Pacific Northwest 2001 weed management handbook*. Corvallis: Oregon State University.

Wilson, D. (1970). Blackcurrant breeding: A progeny test of four cultivars and a study of inbreeding effects. *J. Hort. Sci.* 45:239-247.

Wilson, D. and J. Adam. (1966). The inheritance of some yield components in black currant seedlings. *J. Hort. Sci.* 41:65-72.

Wood, C.A. (1960). Pomology. In *1959-1960 Report of the Scottish Horticultural Institute* (pp. 13-23), Invergowrie, Dundee, Scotland.

Yeager, A.F. (1938). Pollination studies with North Dakota fruits. *Proc. Am. Soc. Hort. Sci.* 35:12-13.

Zatyko, J.M. and F. Sagi. (1972). Currant harvesting made easier by spraying with Ethrel. *Acta-Agron. Academiae Scientiarum Hungaricae* 21(3/4):412-418.

Zazulina, N.A. (1976). Self-fertility and capacity for cross pollination in new varieties of blackcurrant [in Russian]. In *Kratkie Tezisy Dokl. 2-i Vses. Konf. Molodykh. Uchenykh Po Sadovod* (pp. 267-268). Michurinsk, Russia.

Zubeckis, E. (1962). Ascorbic acid content of fruit grown at Vineland, Ontario. In *Report of the Horticultural Experiment Station and Products Laboratory* (pp. 90-96), Vineland, Ontario, Canada.

# Index

Page numbers followed by the letter "t" indicate tables; those followed by the letter "f" indicate figures.

Printed in the United States
by Baker & Taylor Publisher Services